Complexity and Simplicity

Transdisciplinary Studies

Scope

Transdisciplinary Studies is an internationally oriented book series created to generate new theories and practices to extricate transdisciplinary learning and research from the confining discourses of traditional disciplinarities. Within transdisciplinary domains, this series publishes empirically grounded, theoretically sound work that seeks to identify and solve global problems that conventional disciplinary perspectives cannot capture. *Transdisciplinary Studies* seeks to accentuate those aspects of scholarly research which cut across today's learned disciplines in an effort to define the new axiologies and forms of praxis that are transforming contemporary learning. This series intends to promote a new appreciation for transdisciplinary research to audiences that are seeking ways of understanding complex, global problems that many now realize disciplinary perspectives cannot fully address. Teachers, scholars, policy makers, educators and researchers working to address issues in technology studies, education, public finance, discourse studies, professional ethics, political analysis, learning, ecological systems, modern medicine, and other fields clearly are ready to begin investing in transdisciplinary models of research. It is for those many different audiences in these diverse fields that we hope to reach, not merely with topical research, but also through considering new epistemic and ontological foundations for of transdisciplinary research. We hope this series will exemplify the global transformations of education and learning across the disciplines for years to come.

VOLUME 7

Complexity and Simplicity

Edited by

Ewa Błaszczak, Małgorzata Cebrat, Jakub Jernajczyk,
Bartłomiej Skowron, Weronika Urbańska and
Roland Zarzycki

BRILL

LEIDEN | BOSTON

Funding was provided by Wrocław Academic Centre (WCA), Poland.

Cover illustration: The order of entropy 10, 2020. Image by Anna Trojanowska

All chapters in this book have undergone peer review.

The Library of Congress Cataloging-in-Publication Data is available online at https://catalog.loc.gov

Typeface for the Latin, Greek, and Cyrillic scripts: "Brill". See and download: brill.com/brill-typeface.

ISSN 2589-6261
ISBN 978-90-04-71452-6 (paperback)
ISBN 978-90-04-71453-3 (hardback)
ISBN 978-90-04-71454-0 (e-book)
DOI 10.1163/9789004714540

This book is printed on acid-free paper and produced in a sustainable manner.

Contents

Acknowledgements VII
List of Figures and Tables VIII
Notes on Contributors X
List of Authors Including Their Specializations XII

1 Introduction 1
Ewa Błaszczak, Małgorzata Cebrat, Jakub Jernajczyk, Bartłomiej Skowron, Weronika Urbańska and Roland Zarzycki

2 A Dispute over Acting Methods 9
Krzysztof Boczkowski

3 V(D)J Recombination: How Simple Elements Lead to a Complex System for Antigen Recognition 19
Małgorzata Cebrat

4 Degrees of Complexity in the Recognition of Substrates by Proteolytic Enzymes 34
Marcin Drąg

5 Shocking Harmony 43
Magda Dubińska-Magiera and Ewa Błaszczak

6 Can Consonance Be Explained with Mathematical Proportions? The Simplicity and Complexity of Harmonic Relations in Music 59
Adrian Foltyn

7 Reinforcement Learning Tasks: From Complexity to Simplicity 70
Dorota Frydecka and Jarosław Drapała

8 Art: A Particularly Sophisticated Simplicity 86
Łukasz Huculak

9 The Complexity of a Straight Line 96
Jakub Jernajczyk

10 Simplifying Complexity: Copper(I) Complexes with Phosphines Derived from Peptides to Enhance Cancer Treatment 107
Urszula K. Komarnicka and Ewa Błaszczak

11 The Complexity of the New Media 119
Katarzyna Kopecka-Piech

12 The Simplicity and Complexity of Philosophical Discourse: Two Different Ways of Reasoning about the Stoic Postulate of Premeditation of Evils (*Praemeditatio malorum*) 129
Marek Krajewski

13 The Complexity and Simplicity of Zinc Proteins: Minimum Form, Maximum Substance 142
Artur Krężel

14 Why Do We Use Simple Organisms to Model Complex Human Diseases? 148
Marta Migocka-Patrzałek

15 The Beauty and Simplicity of the Complex Biological Chemistry of Metals 161
Magdalena Rowińska-Żyrek

16 On Simplificatory and Diversificatory Aspects of the Presentation of an Object 175
Bartłomiej Skowron

17 The Complexity and Multifacetedness of Conversation 193
Karol Striżyk, Piotr Kruk, Justyna Ziółkowska, Bartłomiej Skowron and Magdalena Witkowicz

18 Algorithmic Truth Theory (ATT) 206
Roland Zarzycki

19 Some Comments Regarding the Digitalisation of the Health Care System in Poland during the COVID-19 State of Epidemic: Selected Legal Aspects 226
Katarzyna M. Zoń

20 Works of Art 236
Contributing Artists

21 Conclusions 273
Ewa Błaszczak, Małgorzata Cebrat, Jakub Jernajczyk, Bartłomiej Skowron, Weronika Urbańska and Roland Zarzycki

Index 279

Acknowledgements

The book "*Complexity and Simplicity*" is the outcome of an original and interdisciplinary undertaking by the Academy of Young Scholars and Artists based in Wrocław, Poland. Both the Authors and Editors would like to express their sincere gratitude to the Wrocław Academic Centre (WCA; Poland) for financial support.

We would also like to thank Athina Dimitriou, Acquisition Editor at Brill, Christine Hededam, Associate Editor for Education at Brill, and, last but not least, Priya Ramesh, Production Editor at Brill, for their invaluable assistance, guidance, and support throughout the publication process.

Akademia Młodych Uczonych i Artystów

10 lat WCA Wrocławskie Centrum Akademickie

Figures and Tables

Figures

3.1 Structure of a T cell receptor (based on TCRαβ) and an antibody. 21
3.2 Course of V(D)J recombination. 22
3.3 Maturation of Tαβ cells in the thymus and the selection of the TCR repertoire. 25
3.4 The scenario of the transposon hypothesis for the origin of *RAG-1* and *RAG-2*. 27
3.5 Characteristic locations of genes originating from transposons that still function in their hosts' genomes. 28
3.6 Structure of the *RAG/NWC* locus. 29
5.1 The types of protein structure. 46
5.2 A simplified diagram of the HSP70 protein's reaction cycle. 48
5.3 A simplified diagram of the network of HSP family proteins, the main component of the protein quality control system (PQC). 49
5.4 A simplified diagram of a protein life cycle. 50
6.1 Geometric representation of simple ratios by Descartes. 62
6.2 Measure of harmonic entropy of order F = 64 for intervals between the unison and the octave + major third. 63
6.3 Pattern window of model responses to two signals with periods of 4 ms and 5 ms (250 Hz and 200 Hz), forming the major third interval. 65
6.4 The F1 score computed for four described measures and experimental results, based on ranking thresholds discriminating dissonances and consonances. 67
7.1 The flow of a single trial in the test. 73
7.2 Description of the neuron model proposed within the Leabra framework. 75
7.3 Conceptual model of *basal ganglia* with corresponding neural model. 77
7.4 Graphical interpretation of formula (4). 80
9.1 W. Gołuch, *Double parabole*, 1979. 97
9.2 W. Gołuch, *Cut out that line*, 2011 (parts 0, 9, 10). 99
9.3 W. Gołuch, *Falling dot*, 1979. 101
9.4 W. Gołuch, *Infinite drawing*, 2011. 104
10.1 Copper(I) complexes with phosphines derived from fluoroquinolones. 110
10.2 Copper(I) complexes with different phosphine ligands. 111
10.3 Schematic route of phosphine ligands with and without peptide, as well as copper(I) complex with phosphine ligands. 112
10.4 Schematic mechanism of cytotoxic action of copper(I) complexes with phosphine-peptide conjugate. 113
14.1 Fruit fly (*Drosophila melanogaster*). 150

14.2 The key moments in mice mutagenesis. 152
15.1 The complex system of nickel homeostasis in *H. pylori*. 164
15.2 A) A scheme of Heat Shock Protein A from *H. pylori*; B) A suggested scheme of the structural rearrangement of HspA upon Bi3+ addition; only the front subunit is highlighted (Cun et al., 2008). 166
15.3 Binding modes of nickel to the Hpn sequences discussed above -HHYH, MAH and CC. 167
15.4 Secondary structures of (A) the genomic HDV ribozyme (B) the CPEB3 ribozyme (C) the small model constructs P1, P2 and P4 used for resonance assignment; the nucleotides added to the natural sequences are shown in grey. 170
15.5 Schematic model of *C. albicans* zinc scavenging from host cells. 172
16.1 The Herman grid illusion. 184
16.2 A series of objects of progressively higher dimensional type. 187

Tables

6.1 Musical intervals and their corresponding simple ratios in just intonation. 61
6.2 Comparison of selected measures of consonance in terms of distinction between consonances and dissonances. 66

Notes on Contributors

Ewa Błaszczak
is Molecular Biologist, currently a University Professor at the Department of Biochemistry and Molecular Biology at the Medical University of Lublin, Poland. From 2016 to 2022, she was a researcher and lecturer at the Faculty of Biological Sciences at the University of Wrocław, Poland. She completed her doctoral thesis at the Institute of Genetics and Development of Rennes (IGDR), France, and obtained a Ph.D. in Biology from the University of Rennes 1 in 2015. Earlier, she graduated from the University of Westminster in London, UK (M.Sc. in Medical Molecular Biology) and the University of Life Sciences in Lublin, Poland (M.Sc. in Biology). Her research interests focus on molecular mechanisms of cancer development, progression and invasion. She is particularly interested in proteins, their interactions and post-translational modifications that influence various cellular processes. Since 2017, she has been a member of the Academy of Young Scholars and Artists, and in the years 2019–2021 she was a member of the Presidium of the Academy. Additionally, she is a philologist and an English translator with a Diploma in Translation Studies obtained from the University of Portsmouth, UK. She participated in the 68th Lindau Nobel Laureate Meeting that took place in Germany. She also popularizes science through workshops and lectures.

Małgorzata Cebrat
is currently Head of the Laboratory of Molecular and Cellular Immunology at the Ludwik Hirszfeld Institute of Immunology and Experimental Therapy, the Polish Academy of Sciences, Wrocław. She is a molecular immunologist, researcher, lecturer, and beekeeper. Her research interests focus on the control of gene expression during the maturation of the immune system and the evolution of the adaptive immune system. Her scientific work centres, moreover, on topics related to the gender of bees, particularly as seen from the perspective of molecular biology and genetics. This includes recently identifying gene variants of CSD that determine the sex of bees. She graduated from the Faculty of Biotechnology at the University of Wrocław and did her Ph.D. in Biology at the Institute of Immunology and Experimental Therapy. In the years 2010–2013, she was the Chairwoman of the Academy of Young Scholars and Artists. She is the author of numerous peer-reviewed publications, and works to popularize science among young people.

Jakub Jernajczyk
is Visual Artist, popularizer of science, Ph.D. in the Arts, MA in Mathematics. He graduated from the Faculty of Mathematics and Computer Science of the University of Wrocław and the Faculty of Graphic Arts and Media Art at the Eugeniusz Geppert Academy of Art and Design in Wrocław. Currently Dean of the Doctoral School and Professor at the Media Art Department at the Academy of Art and Design in Wrocław. He has participated in many scientific conferences and many artistic exhibitions in Poland and abroad (including, but not limited to, Germany, South Korea, Australia, the United States, China, New Zealand, Romania, Hungary, Great Britain, the Czech Republic and Slovakia). In his creative, scientific and didactic work he mainly focuses on the relationship between the arts and science. In particular, he is interested in the part media arts may, and should play, in the popularization of science. He creates installations, most frequently of a multimedia-based and interactive kind, in which he deals creatively with scientific issues.

Bartłomiej Skowron
is Platonist mathematical philosopher. He earned his Ph.D. in Philosophy in 2012, as well as a BSc in Mathematics in 2011, from the University of Wroclaw, Poland. Since October 2015 he has held the position of Assistant Professor at the Faculty of Administration and Social Sciences at the Warsaw University of Technology. He is a member of the International Center for Formal Ontology and the Academy of Young Scholars and Artists, where he served as President from 2013 to 2017. He developed topo-ontology, an ontological framework that posits ideal spatial qualities as fundamental. His current research focuses on the development of a dynamic theory of ideas, in which he defends a contemporary form of Platonism against all kinds of anti-Platonists through the application of category theory. He seeks to further the flourishing of philosophy by both practising it and promoting it through research and educational initiatives, such as *Proper Thinking* (critical thinking workshops for high-school students) and *Filomates* (a project exploring the non-obvious connections between mathematics and philosophy). He has authored over 45 original articles, as well as a book entitled *Part and Whole: Towards Topo-Ontology*. Beyond this, he is an enthusiast of long-distance running and a fan of newly emerging technologies.

Weronika Urbańska
is Doctor of Engineering and Technology in the discipline of Environmental Engineering, Mining and Energy, specializing in Waste Technology; Assistant Professor (undertaking research and teaching) at the Faculty of Environmental

Engineering, Wrocław University of Science and Technology. Her scientific interests include hydrometallurgical and biological methods of recovering metals from waste Li-ion batteries, as well as sustainable raw materials management, Zero Waste and the possibility of recovering metals from space resources. Author of over 60 papers in the field of waste management; participant in national and international scientific and industry-related conferences. In 2021, she was "Climate Person of The Year" in the competition run by the Ministry of Climate and Environment, for her activities directed towards the environment and counteracting climate change. President of the Wrocław Academy of Young Scholars and Artists (2023–2024). Academic tutor and participant of the Masters of Didactics project, as part of which she undertook a study visit to University College London in 2021.

Roland Zarzycki
is holding dual doctorates in Mathematics and Humanities. He serves as Deputy Director of the Gabriel Narutowicz Institute of Political Thought, being also academically affiliated at the Department of Sociology at Collegium Civitas. Academic teacher, researcher, author of scientific publications and speaker at international conferences. Dr. Zarzycki collaborates internationally as a political and economic expert, especially in the areas of global responsibility, social economy, the development of artificial intelligence and the implementation of the Sustainable Development Goals. He has co-founded and coordinated dozens of scientific, social, educational, and cultural projects. His academic interests centre on the critical analysis of fantasy and ideology in socio-political contexts and everyday life.

Authors Including Their Specializations

1. Ewa Błaszczak (health sciences), Department of Biochemistry and Molecular Biology, Faculty of Medical Sciences, Medical University of Lublin, W. Chodźki 1, 20–093 Lublin, Poland; Previous address: Department of Genetics and Cell Physiology, Faculty of Biological Sciences, University of Wroclaw, Kanonia 6/8, 50–328 Wroclaw, Poland.
2. Krzysztof Boczkowski (art: theatre), AST National Academy of Theatre Arts in Krakow, Wrocław Branch, Braniborska 59, 53–680 Wrocław, Poland.
3. Małgorzata Cebrat (biological sciences), Laboratory of Molecular and Cellular Immunology, Department of Tumor Immunology, Institute of Immunology and Experimental Therapy, Polish Academy of Sciences, Weigla 12, 53–114 Wrocław, Poland.
4. Jarosław Drapała (computer sciences), Department of Computer Science and Systems Engineering, Faculty of Information and Communication Technology, Wrocław University of Science and Technology, Wyb. Wyspiańskiego 27, 50–370 Wrocław, Poland.
5. Marcin Drąg (chemistry), Department of Bioorganic Chemistry, Faculty of Chemistry, Wrocław University of Science and Technology, Wyb. Wyspiańskiego 27, 50–370 Wrocław, Poland.
6. Magda Dubińska-Magiera (biological sciences), Department of Animal Developmental Biology, Faculty of Biological Sciences, University of Wrocław, Sienkiewicza 21, 50–335 Wrocław, Poland.
7. Adrian Foltyn (art: music), Institute of Musicology, Faculty of Historical and Pedagogical Sciences, University of Wrocław, Szewska 36, 50–139 Wrocław, Poland.
8. Dorota Frydecka (medicine: psychiatry), Department of Psychiatry, Faculty of Medicine, Wrocław Medical University, Wyb. L. Pasteura 10, 50–367 Wrocław, Poland.
9. Łukasz Huculak (art: painting), Department of Painting, Faculty of Painting, The Eugeniusz Geppert Academy of Art and Design in Wrocław, pl. Polski 3/4, 50–156 Wrocław, Poland.
10. Jakub Jernajczyk (art: visual arts, mathematics), Department of Media Art, Faculty of Graphics and Media Art, The Eugeniusz Geppert Academy of Art and Design in Wrocław, pl. Polski 3/4, 50–156 Wrocław, Poland.
11. Urszula K. Komarnicka (chemistry), Department of Bioinorganic and Biomedical Chemistry, Faculty of Chemistry, University of Wroclaw, Joliot-Curie 14, 50–383 Wroclaw, Poland.

12. Katarzyna Kopecka-Piech (media and communication studies), Department of Mediatization, Institute of Social Communication and Media Studies, Maria Curie-Skłodowska University in Lublin, Głęboka 45, 20–612 Lublin, Poland.
13. Marek Krajewski (art: writing), a Polish crime writer and linguist, former assistant professor and lecturer at the Institute of Classical Philology and Ancient Culture of the University of Wrocław, Poland.
14. Piotr Kruk (psychology), Wrocław, Poland.
15. Artur Krężel (biotechnology), Department of Chemical Biology, Faculty of Biotechnology, University of Wrocław, Joliot-Curie 14a, 50–383 Wrocław, Poland.
16. Marta Migocka-Patrzałek (biological sciences), Department of Animal Developmental Biology, Faculty of Biological Sciences, University of Wrocław, Sienkiewicza 21, 50–335 Wrocław, Poland.
17. Magdalena Rowińska-Żyrek (chemistry), Faculty of Chemistry, University of Wrocław, Joliot-Curie 14, 50–383 Wrocław, Poland.
18. Bartłomiej Skowron (philosophy, mathematics), International Center for Formal Ontology, Faculty of Administration and Social Sciences, Warsaw University of Technology, pl. Politechniki 1, 00–661 Warsaw, Poland.
19. Karol Striżyk (psychology), Wrocław, Poland.
20. Magdalena Witkowicz (psychology), Centre for Interdisciplinary Research into Health and Illness, Faculty of Letters, University of Wrocław, św. Jadwigi 3/4, 50–266 Wrocław, Poland.
21. Roland Zarzycki (mathematics, social sciences), Collegium Civitas, Defilad Square 1, 00–901, Warsaw, Poland.
22. Justyna Ziółkowska (psychology), Department of Humanities, Warsaw School of Social Sciences and Humanities, Wrocław, SWPS University, Chodakowska 19/31, 03–815 Warszawa, Poland.
23. Katarzyna M. Zoń (law), Department of Law, Faculty of Law and Communication in Wrocław, SWPS University, Chodakowska 19/31, 03–815 Warsaw, Poland.

The authors of the artistic works are affiliated with the Eugeniusz Geppert Academy of Art and Design in Wrocław (pl. Polski 3/4, 50–156 Wrocław, Poland).

CHAPTER 1

Introduction

Ewa Błaszczak, Małgorzata Cebrat, Jakub Jernajczyk, Bartłomiej Skowron, Weronika Urbańska and Roland Zarzycki

As one of the protagonists of Clare Boothe Luce's novel notes, "the height of sophistication is simplicity". This dialectical intertwining of simplicity and complexity accompanies us in everyday life when we discover that seemingly simple phenomena hide complex mechanisms, and understanding them requires considerable cognitive effort. The complexity of simplicity is also experienced when we try to synthesise the chaos of phenomena, theories and opinions that surround us. We observe it when the oversimplifications typical of populist discourses lead to tragic political consequences. At the same time, complexity and simplicity can be perceived as aspects of the relationship between human perception and the forms of its expression that constitute social reality. A sign, a thing, a simulacrum appears to us as simple or complex not only due to our cognitive condition but also our emotional, ethical and aesthetic disposition. We note that openness to the complexity of perspectives and simplicity of expression has always been a virtue as well as an ally in our human attempts to understand the world, both on the paths of science and art.

The multidimensionality, diversity and complexity come easier when we are not alone in the cognitive process. Shared reflection and exchange of thoughts support synthetic approaches that allow complexity and simplicity to be reinterpreted over and over again. And this was the main motivation underlying another act of the unique endeavour, namely the interdisciplinary project "Excess and Lack" implemented within the Academy of Young Scholars and Artists in Wrocław in 2013 (Huculak et al., 2013). One of the fundamental objectives of the Academy is to improve and foster communication between outstanding representatives of various fields of science, the humanities and art. Thus the idea behind "Excess and Lack" was to present various perceptions of the concepts of "excess" and "lack" from the perspective of different disciplines in order to reveal similarities, differences and fresh perspectives, thus proving the complementary nature of all these contributions. The importance of the ability to understand a given concept across multiple different languages refers back to the ideas of many great scholars and artists, including Ludwig Wittgenstein's notion of "seeing-as".

In this context *Complexity and Simplicity* should be considered a follow-up and a further step along this path, driven by many enthusiastic responses to the former project. This time, to close the gap between the sciences, humanities and the arts, and replace it with a unifying and holistic perspective on human cognition, 57 scholars and artists were asked to present their perspectives on the concepts of "complexity" and "simplicity" in terms that referred to their own specific disciplines. The essays that emerged from this project are characterised by scientific excellence and anchored in state-of-the-art research, while at the same time being written in a language and style that facilitates their comprehension by an audience new to these specific fields. Also, all authors were asked to present fruitful ideas relating to the broader horizons of their particular disciplines rather than just focusing on their cutting-edge discoveries. Hence, the articles presented in this compilation may serve as an interdisciplinary source of inspiration for a wide variety of readers. These contributions are presented in a complimentary manner alongside works by artists that touch on the issues of simplicity and complexity in the language of art.

Such an interdisciplinary approach seems particularly justified precisely in the case of simplicity and complexity. The search for comprehensive pathways to capture the tension between these two qualities has a long tradition, just to mention Spinozian thought attempting to delineate the chaos within an existing whole and harmony. Although the very issue of complexity and simplicity can be considered within each of the individuated scientific and artistic disciplines, it can and should also become a subject of reflection on a higher level of abstraction. Our book, then, can also be read in these two ways, as an overview of the detailed, diverse grasps of simplicity and complexity, and as a kind of tracing of a common interdisciplinary denominator in order to discover the deeper nature of the title opposition. As with the Deleuzian assemblages, the success of the latter approach requires finding a unique fusion of material interdependencies and constellation of perspectives.

Simultaneously, the paths of reflection quickly branch out into a thicket of parallel intellectual endeavours that the reader may wish to follow, leading us to more contemporary research undertaken in the fields of new materialism, critical realism, but also reflections on the posthuman condition and interactions between the human body, technology and machines by Braidotti, or investigations in heuristic theory of value by Cantó Milà. It is not clear whether adding further crossroads will simplify the matter or make it even more intricate. Nevertheless, this is the risk we have chosen to take in this publication, adding a handful of our own insights to the long-lasting debate on complexity and simplicity.

This unique fusion is presented to the reader in the form of a mosaic of scientific texts and selected works of art. As in Cortazar's *Rayuela*, the reader can choose his or her own path of travel through the meanders of simplicity and complexity by assimilating this publication according to an individually chosen key and sequence. At the same time, in the opinion of the editors of this volume, one of the natural and firmly grounded paths of cognition is to move from the level of detailed scientific investigations concerning the biological, chemical and physical structure of the human being, through theories describing the functioning of individuals, communication theory and the social sciences, to narratives touching on metareflections on the very mode of cognition, expressed many times so successfully through artistic lenses.

Following this approach, the reader may begin this journey from the text delivered by M. Rowińska-Żyrek "The beauty and simplicity of the complex biological chemistry of metals", where the interaction of metal ions with a variety of biomolecules, including proteins, peptides and human ribozymes is described. Despite the sophistication of these phenomena, the author demonstrates in a highly accessible way how simplicity and complexity intersect in modelling these interactions. The article "V(D)J recombination: How simple elements lead to a complex system for antigen recognition" explains, in turn, why flexibility and complexity seen as a feature of the immune system of vertebrates are key factors in their evolutionary success. Importantly, thanks to excellent guidance by the author, M. Cebrat, this fascinating journey through genetic invasions and attempts to model them in the language of modern biochemistry proves comprehensible to the reader. Another aspect of the functioning of the human body at the molecular level is proposed in the paper by A. Krężel "The complexity and simplicity of zinc proteins: Minimum form, maximum substance", in which the author shows us how the differentiation of the kinetic and thermodynamic stability of the zinc complex in proteins depends on the enormous complexity of these macromolecules related to their biological function. Along a natural process, the simplicity of the building blocks is combined here with the complexity of the structure, leading to the creation of an object with extremely intricate properties: the zinc metalloproteome.

The complexity and simplicity of chemical models of human body functioning are also addressed in the text by M. Drąg "Degrees of complexity in the recognition of substrates by proteolytic enzymes." Interestingly, proteases from the exopeptidase family are subject to relatively simple mechanisms, whereas the mechanism of binding and recognition of substrates in the case of endoproteases turns out to be complex. Proteostasis is also the subject of "Shocking harmony" by M. Dubińska-Magiera and E. Błaszczak. As argued by the authors, the preservation of life requires the maintenance of dynamic equilibrium, and it is the harmony of the proteome that is needed to maintain it. This harmony, the so-called proteostasis, observed at the level of a whole set of proteins finds its surprising counterpart at the level of the synchronisation of the cognitive system's responses with the frequency of incoming sounds, as described by A. Foltyn in the text "Can consonance be explained by mathematical proportions? Simplicity and complexity of harmonic relations in music." The very diversity of applications of the concept of harmony combined with the structural parallel between the paths of perception of the world and the paths of its scientific modelling shows us something fascinating about the paths of our human cognition.

In the paper "Why do we use simple organisms to model complex human diseases?" M. Migocka-Patrzałek in a yet different manner tackles the phenomenon of explaining and modelling complex objects by analysing the simpler components on the one hand and the subtleties of structure on the other. The author explains how the use of simple organisms, such as worms or fruit flies, allows us to say so much about the functioning of the much more complex human organism. A medical perspective on the issue of simplicity and complexity can be found in the text by U. K. Komarnicka and E. Błaszczak, "Simplifying complexity: Copper(I) complexes with phosphines derived from peptides as potent anticancer drugs", which addresses the complexity of the effects of anticancer therapies and their relevance to the side effects of contemporary therapies. A different standpoint on the issue of complexity is presented in the paper of K. M. Zoń in "Some Comments Regarding the Digitalisation of the Health Care System in Poland during the COVID-19 State of Epidemic: Selected Legal Aspects", where the legal dimension of the digitalisation of the healthcare system in Poland is described. The increasing complexity of science itself has serious practical and material consequences and translates into the development of administrative processes and legal solutions organising human cognition, which is of considerable importance.

In the text "Reinforcement learning tasks: From complexity to simplicity" the reader encounters an intriguing fusion of computer science and psychiatry specific to the field of computational psychiatry. D. Frydecka and J. Drapała discuss how the behaviour of the human brain varies during decision-making, where some processes have a surprisingly simple explanation, while others require complex modelling. The complexity of signals in a world of information overload, this communication jitteriness, leads to confusion, which, at the level of socially diagnosed aspirations and attitudes, points to a need for simplicity, more and more frequently expressed by individuals. On the other hand, the simplest processes, such as everyday conversations, turn out to be places where complexity creeps in. An interesting study of this issue is presented in the text by K. Striżyk, P. Kruk, J. Ziółkowska, B. Skowron, and M. Witkowicz "The complexity and multifacetedness of conversation." Paradoxically, human attempts to escape complexity can fuel its presence. On the borderline between linguistics and philosophy, we find a text by M. Krajewski "The simplicity and complexity of philosophical discourse: Two different ways of reasoning about the Stoic postulate of premeditation of evils (*praemeditatio malorum*)", where the author shows how decisive the question of interpreting words and their meaning is. And so the very understanding of simplicity, as considered from a literary-persuasive and analytical perspective, is of great importance for our life and cognitive choices.

Eventually, an ontological analysis of the phenomenon of simplicity and complexity itself is provided by the text B. Skowron "On simplificatory and diversificatory aspects of the presentation of an object", where the author makes use of the concepts of K. Twardowski, who is well known in Poland as the master of masters of philosophy. Structure complexity analysed from a contemporary communicative perspective becomes the subject of K. Kopecka-Piech's paper "The complexity of the new media." From this analysis, we learn how the complexity inherent in the culture of our thinking and communication ultimately translates into our perception of the world. In the text "Algorithmic Truth Theory (ATT)," R. Zarzycki presents the challenges of the very process of cognition and points to the growing information complexity, in which a lost individual begins to seek a qualitatively new criterion of truth.

K. Boczkowski takes us to another dimension of the phenomenon of complexity, by addressing the topic of the performative exhibition of the world in the space of the theatre. The complementarity and tension between complexity and simplicity present in the concepts of acting are discussed in his text "A dispute over acting methods." The subtle game of complexity is described in the text "A particularly sophisticated simplicity" by Ł. Huculak, in which the author invites the reader to a fascinating journey through the dialectic of simplicity and complexity, illustrated with vivid and unique insights of artists from past epochs. Another journey, no less interesting, is proposed by J. Jernajczyk in the text "The complexity of a straight line", where he combines mathematical modelling and visual imagination to approach the main concept of this volume from yet another angle.

The second path through the book is marked by reproductions of pieces of art addressing the dichotomy of complexity and simplicity by purely artistic means, at the same time reflecting its geometric, topological, statistical, semantic, cultural, social, aesthetical, political, and philosophical aspects, not always expressible by verbal means. The authors of these works, coming from the Wrocław artistic community (mainly from the Eugeniusz Geppert

Academy of Art and Design in Wrocław), represent various subdisciplines of the fine arts, such as painting, sculpture, graphic design, print making, media art, ceramics, glass, and design. In this sense, although the path of science and the path of art touch on the same phenomena in a complementary way, they are mutually incompatible.

Often, a work of art can convey an idea in a much more simple way than even the longest book. This does not, however, mean that it is not complex. Following Gadamer, a work of art can be perceived through the prism of a complex game between the author and the recipient, who, through an extraordinary fusion of horizons, becomes its co-creator. If, in accordance with chaos theory, we consider that everything that is past (or at least everything within a certain light cone) influences a given work of art, and on the other hand, a given work of art inspires everything that is future (within another light cone) in a similar way, it turns out that simplicity reveals infinite layers of complexity reconstructable in relation to the continuum of contexts extending arbitrarily far back and into the future.

This non-verbal journey through complexity and simplicity consists of several works made using the acrylic on canvas technique, which present intriguing painterly insights into the essence of complexity and simplicity: *Cordyceps* by Ł. Huculak, *The Power of a Seemingly Simple Interior* by A. Kołodziejczyk as well as *Adoratio* by J. Baśnik. Each of these works illuminates certain aspects of complexity through a subtle play with the means of artistic expression. We observe another interesting instance of this play in the lithograph on marble *SHADES_07* by A. Trojanowska.

The interpenetration of the complexity of form and the stubborn simplicity of the material is manifested in the glass works by A. Joszczuk-Brzozowska (*Materialization*), by B. Mak-Sobota (*Onestring*), and A. Leśniak-Banasiak (*Amoria*). The austerity of form is also a feature of the slumped glass and metal composition *The Stijl* by M. Krzemińska-Baluch. A disturbing disruption to the seemingly simple form is proposed by the works *Proximity I, II* by B. Sacharczuk, made using the stoneware moulded on a potter's wheel technique. The expression of the works of A. Skowroński, *Wrocław Sienkiewicza* (digital print), and M. Mierzicki, *PLAY MY BOY* (object, steel) seem to be particularly efficient, achieving a seemingly outstanding depth of political and existential interpretations, despite the simplicity of the form.

D. Sobolewska's exhibition *RGB* exposes the viewer to an individual encounter with complexity through contact with simple installations. This spatial arrangement of nine cubes made of transparent plastic in the basic additive colours was carried out in cooperation with WRO Art Center. Also K. Freino's work *Cataract* conducts public space intervention to highlight yet another aspect of the tension between simplicity and complexity. A performative approach to the phenomena can also be found in the printmaking performance by Z. Dyrda, *I'm Just a Pawn*, where the meaning of complexity is related to the established game code, in this case, defined by the rules of chess. What makes a pawn just a pawn? I. Juarez touches on simplicity and complexity in an exposition of the series of wooden elements under the title *Insect City*, while M. Grzybowska uses steel and textile rubber in her *Sketches for space*. The issue of symmetries, expansions and dimensionality of the space can be found in the oil-on-canvas painting *Third Diagonal Symmetry* by K. Moskowczenko as well as a drawing by J. Grulkowski, *The Expansion*, but also in the mechanistic discussion with the recipient initiated by P. Jędrzejewski by his *installation 67/68*.

Reflection on simplicity and complexity using more anthropomorphic means of expression can be found in the pieces of art *Priestess* by P. Baśnik (oil on canvas) and *Wonder emptiness – I'm waiting for you* by M. Gorzelak (intaglio, relief printing). A simple aesthetic provocation exposing the recipient to his/her utilitarian experience has been offered by K. Szymanowska in *Sensitive Point*. Words contrasted with artistic expression are to be found in the proposal of E. Smoliński, *On the matter of simplicity and complexity*, where the functional aspects of a folding ruler are being challenged. The issue of measurement also appears in the work of T. Dobiszewski (*Tracking the Sun*), where the complexity of the movement of the sun is captured in the simplicity of the equipment aimed at measuring it. The play with the simplicity of the form and shadow as an element of artistic expression is proposed by J. Żak in the work *Written by shadow* (Pâte de verre).

In *The Great Gatsby* P. Kmita addresses the issue of complexity emerging from the composition of simple elements, namely shots taken from the famous movie under the same title. Recognizing a repeatable and reproduced pattern creates a kind of tunnel between the world of complexity and the world of simplicity, showing that neither complexity nor simplicity are features of reality but of our cognition. A similar discussion seems to be undertaken by M. Kociński, who used the ceramic mosaic technique to create *Shield 7*, and M. Grzyb in *Stories*, where the complexity of human biographies becomes trapped by the tyranny of simplicity of form. The arrangements of small ornaments create complexity, which is then tamed by the recognition of the pattern.

A synesthetic proposition of K. Marusińska (*Cacophony*) deals with complexity expressed as communication chaos emerging from ceramic, acrylic, and resin forms. The complexity in the work by U. Lucińska and M. Knychaus (*Diaphane*) is reflected by a multi-layered pattern made of stainless steel and brass. *Plus minus* by W. Gołuch is based on a dialogue between the simplicity of the sign and the complexity of the meaning. J. Jernajczyk presents *The*

Knot of Cuts, which prompts reflection on the representability of lower-dimensional (simpler) objects by higher-dimensional (more complex) ones. Finally, the image of excess (some kind of complexity) can be found in the work of J. Opalska-Brzecka.

Although it is impossible to summarise the complexity of artistic expressions in a simple introductory description, we believe that at the end of the quest the reader will be able to recollect the common core of the artistic and scientific reflection, which we tried to preserve throughout the volume. Simplicity and complexity do not seem to be either better or worse, in the way they accompany us on our cognitive journeys. In the modern world, filled with cheap sensations, where ignorance turns difference into a threat, and digits replace emotions, we feel that in the common-day sense, pluralism proves its value. And it is this pluralism in perceiving the world that we would like to present to the reader once again to save us all from blaming simplicity for primitivism and complexity for sophistication.

Reference

1 Huculak Ł., Skowron B., Dąbrowska K., Jernajczyk J., Zakrzewska G., & Zarzycki R. (Eds.). (2013). *Excess and lack (Nadmiar i Brak in* PL*), The Academy of Young Scholars and Artists, Wrocław 2013*. http://www.huculak.pl/EN/katalogi/wspolne/brak.pdf

CHAPTER 2

A Dispute over Acting Methods

Krzysztof Boczkowski

Abstract

This paper is a comparative study of two completely different views regarding onstage acting methods. The *Stanislavsky system* has been subject to criticism by David Mamet, who rejects all the main tenets of the method proposed by Stanislavsky. The complexity of Stanislavsky's *method* is juxtaposed with the simplicity of Mamet's conception, which assumes that all that is needed for an actor to act on a theatrical stage is the character's goal, action, or simple onstage situations, as well as openness to their own emotional state and dialogue with their stage partner. Analysis of those two views reveals the direction in which the actor should proceed on stage, starting from a simple goal and action and moving towards the complexity of emotional experience.

Keywords

Konstantin Stanislavsky – David Mamet – Stanislavsky system – the magic if – character – dialogue – partner – action – emotions – broadened mind – organicism

1 Realist Theatre as the Plane of the Dispute

The clash between Konstantin Stanislavsky's and David Mamet's views on the process of creating a role and applying it on stage has brought about a dispute over the best acting method. It is a dispute over theatrical work conducted in a convention known as psychological realism, or realist theatre, which is characterised by a logicality of events, linearity, cause and effect, *mimesis* (imitation, mimicking) and *diegesis* (storytelling). This type of theatre is known as "dramatic theatre" and encompasses the story, stage action, the characters, the utter domination of the text, etc. Perhaps in the current theatrical reality, where postmodernism and avant-garde are commonplace, a more accurate term to use would be "classical dramatic theatre." In this theatrical convention the actor's goal is to accurately achieve the playwright's goals as interpreted by the director and materialised in the course of rehearsals in the form of stage characters played for the audience. The role and theatrical shape of the play performed on stage should be characterised by a broadly understood realism (or even naturalism) and as such represent everyday situations and experiences as closely as possible, or – to use Aristotle's notion of *mimesis praxeos* – show people in action by means of imitation.

The dispute in question is born out of two entirely different views on acting, which may be characterised as a dispute between complexity and simplicity. The simplicity of *being in action* proposed by Mamet is juxtaposed with the complexity of Stanislavsky's *emotional experiencing.* This dispute lets us look at acting from two perspectives and consequently allows us to reject all that is irrelevant and look at the essence of acting.

2 Complexity: Key Notions in the *Stanislavsky System*

To reach an objective we need a set of acting tools, i.e., a method allowing us to achieve channelled and precise acting. The *Stanislavsky system* provides such a set of tools in which one element complements another, producing a coherent whole aimed at achieving a transformation through the creative process. The system's structure is very complex and can be divided into various phases:

- the analytical phase, in which the director and actors "break down" the contents of the play in order to get to know every detail of the character's lives and circumstances as intended by the author and to establish the *supreme objective*;
- the technique phase, which encompasses exercises for improving muscle relaxation, body flexibility, contact with a partner, attentiveness on stage, improvisations about given subjects, activating the actors' imagination, etc., as well as situational rehearsals during which actors try to apply the analytical material gathered and evoke suitable emotions and feelings, i.e., discover *the life of the human spirit on stage* by applying the so-called *magical if,* i.e., putting themselves in their character's shoes and fully believing in that character. Consequently, they maintain the appropriate onstage condition, the same as in offstage life, in order to create realistic situations;
- the staging phase, in which the actors focus on rehearsing the entire play, their characters' emotional experiences, playing the role in a defined personality frame, costume, makeup, internal tempo-rhythm, etc.

Stanislavsky's legacy is a direction, in which the actor should head to create a stage persona that is not a separate "artificial, mechanical being" but the outcome of the actor's onstage emotional experiences, a certain extension of the personality of the actor who finds in himself or herself the required qualities and takes them out into the open while remaining himself or herself. Stanislavsky made it clear that the core idea of his system is the achievement of complete organicism of the actor, succumbing to internal impulses, as well as emotional and external reflexes resulting from the onstage behaviour of their partner.

The phrase *organic body-mind* captures the essence of Stanislavsky's explorations. "The organic body-mind is a body which responds to the demands made by the mind in a way that is neither "redundant," "defaulting" nor "incoherent," that is when:

- the body responds only to the demands made by the mind;
- the body responds to all the demands of the mind;
- reacting to those demands only, the body adapts to them, seeks to satisfy them (Barba & Savarese, 2005, p. 99).

The organic body-mind is "the most normal human condition", the kind we can observe in everyday life. However, the artificial conditions on stage restrict our perception, block it and hinder *the emotional experience* (*perezhivanie*). This most normal human condition, everyday life kind, becomes distorted. What distorts it is our fear of being judged, disbelief in what I am doing, a superficial reception of the play, imprecise action due to stage fright, ignorance and consequently a lack of *suitable* feelings and emotions. Obviously, the actor continues to experience all kinds of feelings and external stimuli. They are, however, remote from the author's and director's intentions, thus to achieve the relevant emotional state of the character, we should use a subtle assortment of psychophysical activities developed during rehearsals. This creates an illusion of everyday life where the artificiality of the stage does not constrain the actors or block out their reactions. The actor becomes the character, both physically and spiritually. He/she believes in what is happening on stage and what they feel as if it were happening in real life. The stage action is purposeful with nothing left to chance and the actors' actions are precise, resulting in a narrative, which makes the audience forget that they are watching a carefully rehearsed performance, leading them to believe in the authenticity of the play. The precise physical actions discovered by the actors enable them to harness the contents coming directly from their subconscious to their acting. The actor's acting is no longer volitional and becomes more involuntary. This has an effect on the actor. The actor surrenders to it as a dancer surrenders to music or a canoeist to the current of the river. However, we are not dealing here with any kind of trans-technique. Emotional experiencing sharpens the senses and harmonises the physical and spiritual aspects of ourselves. The emotional experiencing causes us to perceive reality the way we do in real life, without the mental barriers caused by being on stage in front of an audience.

Stanislavsky's explorations were to transform acting, which had earlier been perceived as make-believe, into something truthful.

3 Simplicity: Mamet Speaks Out against Stanislavsky

David Mamet (2014), who passionately rejects Stanislavsky's ideas is on the opposite end of the acting dispute. He completely negates the need for analysing the text and claims that the knowledge coming out of it is useless to the actor and can only be an obstacle and barrier to a spontaneous being on stage. The actor does not have to 'become' the character, in fact, Mamet does not acknowledge the very existence of the term "character". *A character is only letters on paper.* The actors must speak their lines clearly to achieve more or less what the author intended, and the audience will view them as 'an illusion of the character' on stage. Mamet dismisses emotional experiencing 'of the character's tragedy on stage' and claims that it is impossible to make oneself experience any other feelings than those that he/she has at the moment of being on stage in front of the audience. He ridicules Stanislavsky's method which has the actor think of a memory, reaction or feeling for each moment of the play and remain faithful to it on stage. He claims that such behaviour cuts off the actor's stage awareness and spontaneity. Instead of engaging in dialogues with their onstage partner, the actors are absorbed by checking whether they have activated the emotion "prepared" beforehand. Mamet asserts that the emotional and psychological nature causes a person to rebel against any instruction to experience a feeling. "If a person could really command its subconscious to evoke certain emotions when he/she so wishes, there would be no neuroses, psychoses, psychoanalysis – there would be no sadness" (Mamet, 2014, pp. 24–25). Since we cannot control our thoughts, can we control our emotions?

Stanislavsky's method instructs the actor to think up vicarious stories, called emotional experiencing, with which the actors torture themselves in an attempt to program their mind to experience those emotions. In Mamet's opinion, the actor does not need to experience anything – all they need to do is appear on stage and open their mind and spirit to what happens there, which will trigger the truest and most honest emotions. The actors have their own tragedies. Actors are filled to the brim with stories from their own lives, which make up the emotional load needed to create the role. Stanislavsky's *magical if*, i.e. "what I would do in that situation" (Mamet, 2014, p. 22) is an empty and pointless deliberation because the very idea of a hypothetical situation – such as one that requires frenzied courage or overcoming oneself, as in Hamlet – must be rejected. We cannot foresee what we would do if our mother together with our uncle murdered our father in order to take over the kingdom. "Only a fool or a liar can claim he knows what he would do in circumstances requiring courageous behaviour" (Mamet, 2014, p. 50).

It is the director who builds the meaning of the staging through context and it is he/she who is responsible for structuring the staging in such a way as to provoke the matters included in the play. The actor should under no circumstances be bothered with the meaning of his/her acting and the issues that it conveys. The context and the order of scenes, as well as the simple aim that each scene serves, make up a score that the actor should follow. Let us take a look at a simple example: an actress is sitting on a chair on stage. For the audience, this may mean waiting. The actress is sitting on a chair, holding a watering can and in front of her there is a flower pot with flowers. For the audience, this may mean the intention to water the flowers. The actress is sitting on a chair and in front of her hangs a noose. For the audience, this may mean hesitation before committing suicide. In all the above cases the actress's behaviour is the same. She is relaxed and contemplative while sitting on the chair. She is not trying to imagine or feel anything, she responds to the circumstances of the empty stage, watering can, flower pot and noose. In all those cases she feels the same way. It is the audience who, using their imagination, will make a different interpretation of each of those situations. They do not need to be helped with artificially programmed emotional experiencing. Acting – that fleeting moment – happens when the actors accept all that is going on inside them and around them and combine their own struggles with uncertainty, with the picture sketched by the playwright in the form of the dialogue spoken at that exceptional moment. This creates an illusion of the character, which will be authentic for the audience. The experience of being on stage in front of the audience and the acceptance of that experience is enough to experience the character. It carries a sufficient emotional load.

Mamet compares the actor to an athlete who should have a relaxed, supple and trained body and the courage to overcome obstacles (Mamet, 2014, p. 36). Those obstacles are found on the stage. The most prominent one is our own fear that fills us at the moment we step out of the wings onto the stage. Acting has characteristics of heroism. An actor needs courage, concentration and self-acceptance for the creative moment to take place on stage in front of the audience. The actors, whether they like it or not, stand naked in front of the audience. They do not need to hide behind magic formulae, the character's experiences and history, redundant ideology or ideas for the role. Once the actors accept themselves, they will not fake anything or impose anything on themselves and will open up to the flow of the moment, true acting will become possible.

In summary, Mamet proposes the following: learn the lines, prepare simple onstage actions together with the director (rehearsals should not take longer than a few weeks), set the simplest possible goal to achieve, go on stage and accept your current condition as well as fear and uncertainty, overcome that fear with courage and speak your lines to your partner loudly, clearly and with determination. Ignore your emotions, do not force yourself to feel or imagine anything, be constantly focused on your stage partner. The actor should not hold on to any emotional stirs and fantasies but should resolutely work towards the goal he/she had set. Do not prepare anything redundant, surrender to impulses, be open and humble. That is all. Is it a lot?

Mamet reduces the acting method to a few simple tenets in order to capture its most crucial elements. For that reason he rejects the complexity of Stanislavsky's system, ignoring the fact that at the end of his life, Stanislavsky did a similar thing. He combined all his previous ideas in a new, basic one, which he called *the method of physical actions*, which was to crown his previous explorations. He claimed that there is no system that the working name 'system' embraces only a few basic rules and exercises (see above) and there can be no single recipe for playing a role. The method of physical actions teaches that all the processes occurring in the actor – starting from breathing and going all the way to subtle movements of imagination – come down to exactly that – physical actions. Why, then, does Mamet not discuss the last stage of Stanislavsky's research? Because it is not commonly known and sufficiently described. Mamet makes references to general and superficial opinions about Stanislavsky's work, most of which are inaccurate, erroneous or downright false. Mamet makes numerous remarks about acting being relatively simple while it is academics and all kinds of teachers who come up with quack ideas based on the Stanislavsky system in order to sustain the illusion of possessing "covert knowledge". Mamet tries to dispel that illusion and believes common sense will allow actors to get rid of complexes, making acting simple, comprehensible and tangible. Is it really the case?

4 Acting Practice

Our deliberations are useful only inasmuch as we can identify their practical aspect. Both Mamet and Stanislavsky are in agreement as to the objective of the action and the action itself. Without them, it is impossible to even initiate acting. The key difference lies in the onstage condition. 'The natural human condition' is the stage condition that Stanislavsky recommends to actors. Everything here is subordinated to the truth. Master your technique to the point where the audience does not see acting, by human emotional experiencing, as they do in daily life. However, Mamet encourages the actors to accept their condition of "today", of the "here and now", to let go of any hypothetical experiences of the character. This leads us outside the dramatic theatre into the realm of modern *post-dramatic* theatre. Hans-Thies Lehman (2009) defines this means of theatrical expression as *an invasion of reality*. What has heretofore been a secret, hidden from the audiences' eyes – what the actor feels privately – can become the pivotal point of acting due to the invasion of reality. The line between reality and fiction becomes blurred. Mamet does not notice the trap that the actor may fall into. By accepting their private emotions, they leave their acting to chance. What if the actors feel bored or discouraged, what will be their source of energy to act? What if the dramatic action requires joy and the actors are immersed in their private sadness? By rejecting work on the emotional experiences of the characters, the actor may lose control over the events on stage. Let us try to paraphrase Mamet – what remains is the purpose of the action, the action itself and the stage partner. Focus your attention and involvement on what you are doing and not on how you feel. Who cares about how you feel? Your task is to convey the contents of the play through your actions. Action is what matters the most!

5 From the Simplicity of Action to the Complexity of Emotional Experiencing: An Attempt at Implementing Mamet's Tenets

In 2013, together with two students who were members of the Academic Association at the Acting Division of the fourth year of the Wrocław Branch of the Kraków Theatre School, we did research on *Stage Condition in Daily Situations*. We tried to utilise Mamet's simple tenets. The results of our research were shown in a stage play we titled *An Everyday Situation. O.* The play was based on David Mamet's text titled *Oleanna* (1993; hence the O in the title). The project took about six months. The students focussed only on memorising their lines. We forewent the analysis and only signaled such simple physical actions as: coming up to a locker, taking out books, packing a briefcase, coming up to the window, moving a chair, walking out through the door, etc. Our place of work was a lecture hall with a desk and a board in the middle, to give as much credence to the scene from the play, in which a student meets her professor. Additionally, we wanted to do away with all the theatricality of the place. The actors were asked to completely forego their search for any emotionality or oral interpretation of the text, their internal reactions and the purpose of their actions. Instead, they were instructed to focus on speaking their lines in the simplest manner possible (in theatre jargon, "in white"), in a neutral rhythm while performing the aforementioned actions. Our endeavours were accompanied by loose conversations about the themes of the play. Therefore, a rehearsal would look like this: text work – repeating lines while performing simple actions, intertwined by conversations about Mamet's play. The actors did not attempt to act. We did not try to figure out whether in a given moment we needed anger or sarcasm, request or command, joy or sadness. We did not assume one interpretation of the play and remained open to various interpretations. Neither did we set any goals for our actions, in order to free up our endeavours from an imposed direction.

The effects of our work came as a surprise, because feelings began to emerge spontaneously in the actors. The pace of our conversations became dynamic and changeable. Additionally, clear intentions started to emerge and the actors actually transformed into the characters from the play. The preparations came down to buying costumes (ordinary casual clothes) in a local mall. Did Mamet's assumptions work? In my opinion, this simple approach to acting is not sufficient. It is not enough to arrange "choreographed" moves to replace the complex process of emotional experiencing. It takes more than that, namely the conversations we had during rehearsals, which took more than half of the time. The rehearsals resembled a social event, a loose exchange of opinions rather than the 'table work' on text practised commonly in many theatres. After exhausting subjects contained in the play, we swiftly moved on to personal issues – and when done with those – we went home without feeling guilty about wasting time on "useless talk". What purpose do these conversations serve? Firstly, to build an atmosphere of trust and freedom and remove all obstacles which may stand in the way of the creative act. Secondly, to create a mental plane for the emergence of the *broadened mind*. Franco Ruffi (in: Barba and Savarese, 2005, pp. 162–165) uses this term to describe an actor who – by the conscious application of Stanislavsky's method – begins to experience the play. This happens as a consequence of arousing or awakening the mind through questions and answers, enquiries and fantasies, rejecting and accepting views, approval and negation and the *magical if* (Barba and Savarese, 2005). What would happen if somebody treated me like the character from the play? What would I do if I were to face the same dilemma? The process of looking for the answer is far more important that the answer itself. It is thanks to this process that cool opinions transform into intense feelings, dry definitions become heated emotions. The energy coming from this process leads to involvement. This process gives rise to internal emotional experiencing. When

– as in platonic dialogues – the actor surprises himself/herself with questions for which there is not one answer or it is very difficult to find. They escape clichés and stereotypes in their way of thinking. They genuinely endeavour to get to the bottom of the character's behaviour. What takes place is an unnoticeable process of identifying with the role. Every day the mind becomes filled with the author's ideas. These ideas are no longer distant words on paper. They become the actors' feelings, their life and eventually the actors themselves.

6 Conclusions

In order for drama to occur in front of the audience, the stage partners should – within reason – interact with each other and provoke each other. While analysing the conclusions of our work, the actors said things like: *I really wanted to hit him; I really hated you at times; Your attitude towards me was really disgusting.* The transformation became a fact. The actors did not say: *My character felt this;* or: *I felt it while I was acting.* They used the first person *I* and thus identified their emotions with the emotions of their character. Mamet is right about many things. Building stage action requires the simplest approach possible. A painstaking analysis of the play in terms of the characters' emotions is often unnecessary. One does not need to bring on "faith", which has more to do with tradition and social conditioning than our will. There is no need to plan every detail of being on stage. Neither is the past of the stage character an important issue for the actor to consider. The content overload that directors often throw at us may hinder the creative process. However, this process does need "intellectual fuel". Without it, one finds it hard to activate one's imagination and consequently to experience emotions on stage. Experiencing emotions is not equal to exhibitionism. The actor's reactions are always a priority. They must, however, be related to the play and it is from them that the actor derives impulses that activate the imagination and that become "the motor of emotions". The actor's imagination gives rise to the emotionality of his/her character. It is this emotionality that lifts the barrier between the character and the actor. There is a person and his/her feelings and they are the essence of everything that happens on stage.

References

Barba, E., & Savarese, N. (2005). *Sekretna sztuka aktora.* Wrocław: Ośrodek Badań Twórczości Jarzego Grotowskiego i Poszukiwań Teatralno–Kulturowych.

Gorczakow, M. (1957). *Lekcje reżyserskie Stanisławskiego*. Warszawa: Państwowy Instytut Wydawniczy.

Lechman, H. (2009). *Teatr Postdramatyczny*. Kraków: Księgarnia Akademicka.

Mamet, D. (2014). *Prawda i Fałsz. Herezja i zdrowy rozsadek w aktorstwie*. Myślenice: Wydawnictwo Filmowe.

Stanisławski, K. (2010). *Praca aktora nad sobą. Cześć 1 i 2*. Kraków: PWST im. L. Solskiego w Krakowie.

CHAPTER 3

V(D)J Recombination

How Simple Elements Lead to a Complex System for Antigen Recognition

Małgorzata Cebrat

Abstract

One of the most fascinating features of the vertebrate immune system is its ability to combat never-before-known "aliens". This complex process is made simple by the creation of new gene variants that encode antibodies and T-lymphocyte receptors in B and T lymphocytes. These variants are generated by randomly cutting and reassembling specific DNA fragments, allowing the immune system to potentially recognize an almost limitless array of foreign molecules. This randomness, however, can lead to errors, so the receptors undergo stringent testing to ensure they can identify threats while ignoring the body's own cells. Interestingly, vertebrates gained the ability to cleave DNA to produce these receptors by "infecting" their genome with foreign genes (RAGs). These genes were later adapted to help combat other "aliens". We believe that the NWC gene, a host gene located at the site of RAG gene invasion, played a crucial role in simplifying this genetic assimilation process despite its complexity.

Keywords

immune system – adaptive immune system – antigen recognition – V(D)J recombination – T-cell receptor – *RAG1* – *RAG2*

1 Introduction

The key to a properly functioning immune system is its ability to reliably distinguish between its own and alien elements. The nonspecific response of the immune system involves the recognition of the characteristic parts of invading pathogens and forms the body's first line of defence against infection. The constantly evolving relationship between the host and the pathogen and the resulting on-going change in the structure of pathogens that could be recognised by the immune system often makes a non-specific response insufficient. Thus, the immune system faces a considerable issue: it must recognise something that the body has never encountered before and prepare itself for pathogens that will appear in the biosphere in the future. It seems that the only possible solution is to prepare in advance a vast range of recognising elements (receptors) that could later serve to recognise alien structures (non-self antigens), even those that have not yet evolved. Since receptors are made of proteins, and proteins are coded by genes in the genome, the genome (with its tens of thousands of genes) would be too small to code enough receptor protein variants. Hence, a vast diversity of receptors can be created through the process of V(D)J recombination. The process draws from a limited number of *simple* constituents to create a *complex* system for antigen recognition. Thus, V(D)J recombination forms the basis for the functioning of an adaptive immune system, which is characteristic of vertebrates. The aim of this chapter is to follow the stages of V(D)J recombination, the subsequent selection of receptors and the evolutionary processes that may have led to the development of this seemingly simple yet complex system.

2 Recombination of the Genes of T and B Cell Receptors

The cells that perform a key role in the functioning of the adaptive immune system are T and B lymphocytes. Generally speaking, T and B cells are responsible for the body's second line of defence. They respond precisely to particular antigens by producing antibodies, creating substances that activate other types of cells to combat the invading pathogens and directly killing virus-infected cells and tumorous cells. An important feature of the adaptive immune system is its immunological memory, which enables a quick response once the body comes in repeated contact with a given antigen. B and T cells produce antibodies and TCR molecules, respectively, i.e. antigen receptors (Figure 3.1). Both types of cells consist of the constant domain, which anchors the receptor in the cellular membrane, interacts with other molecules and transmits signals into the cell, and the variable domain, which is responsible for the specific recognition of antigens. A key characteristic of T and B cells is that each cell produces only one type of receptor for a specific antigen. Therefore, the diversity of antigen receptors is due to a large number of lymphocytes, each with a unique receptor.

The variable parts of protein chains in antibodies (Ig) and TCR molecules are coded in the genome of a foetus by noncontinuous gene segments which appear in many variants, called V, D and J minigenes. Genes encoding the

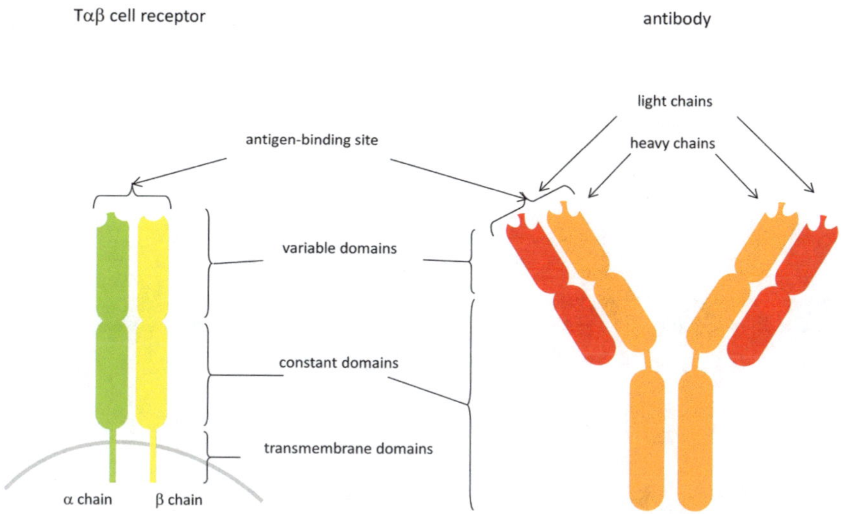

FIGURE 3.1 Structure of a T cell receptor (based on TCRαβ) and an antibody

variable domains of Ig (κ or λ) and TCR (α and γ) light chains are coded by one V minigene and one J minigene, and the variable domains of Ig and TCR (β and δ) heavy chains are coded with one V minigene, one D minigene and one J minigene. In most vertebrate species, the minigenes of each type (V, D or J) are grouped together, with each group neighbouring one another (Bengtén et al., 2000; Litman et al., 1999). For a functional receptor-coding gene to be created, the individual minigenes must become connected in a process called V(D)J recombination, which takes place during the maturation of T and B lymphocytes (Bassing et al., 2002; Tonegawa, 1983). The receptors created through V(D)J recombination owe their variation to combinatorial diversity, which results from the random selection of V, D and J minigene variants, and to joint diversity, which results from the insertion of additional nucleotides when the selected minigenes are joined together (see below) (Lewis, 1994). Each minigene is surrounded by signal sequences composed of conserved parts divided by a 12- or 23-nucleotide spacer (Ramsden et al., 1994; Sakano et al., 1979). Each group of minigenes has a spacer of a characteristic length. For instance, in the locus coding the Ig heavy chain (IgH) in mice, the V minigenes contain sequences with a 23-nucleotide spacer, D minigenes contain a 12-nucleotide spacer and J minigenes contain a 23-nucleotide spacer. During recombination, only minigenes with different lengths of signal sequences join together. This prevents minigenes belonging to the same family (e.g., two or more Vs) from joining together, and also prevents minigenes from joining together in

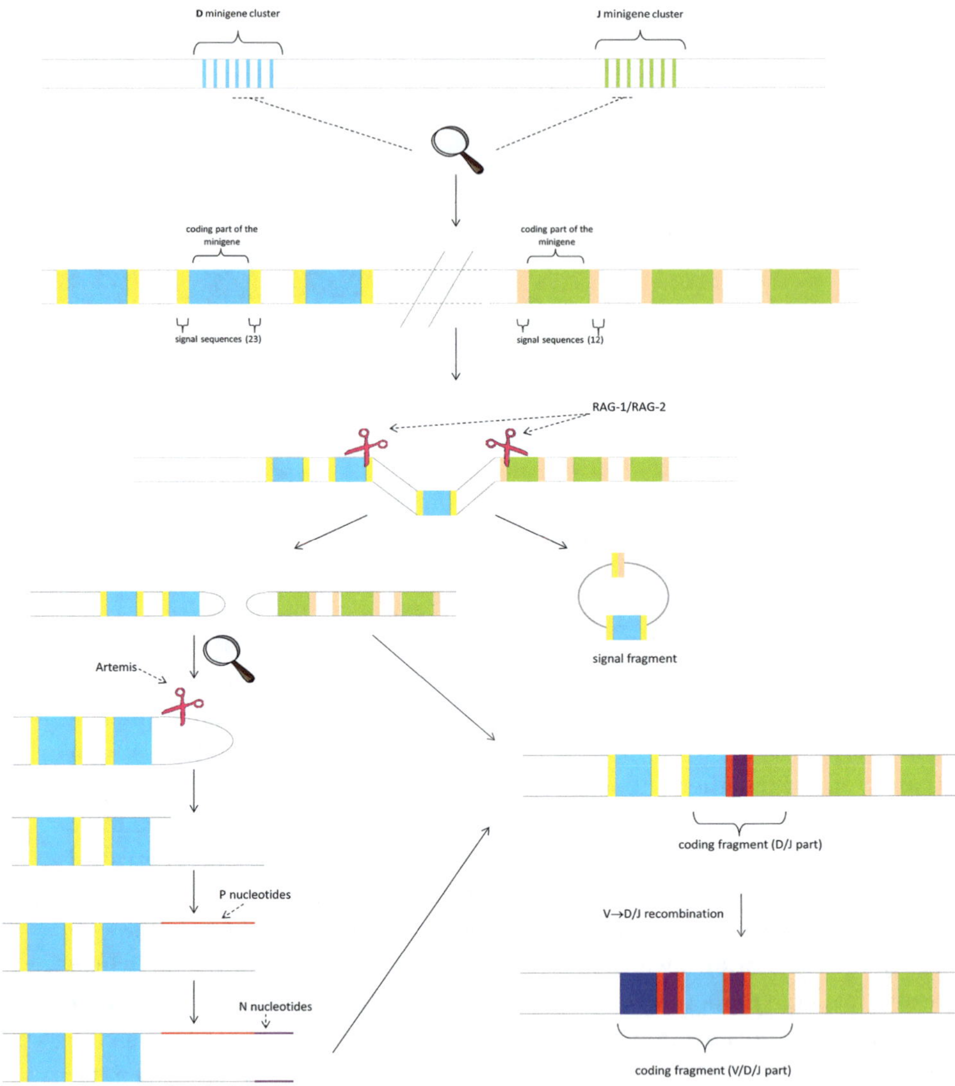

FIGURE 3.2 Course of V(D)J recombination

the wrong order (e.g., a V and a J without a D) (Eastman et al., 1996; Ramsden et al., 1994).

V(D)J recombination (Figure 3.2) can be divided into two phases: the DNA cleavage phase and the DNA repair phase. In the first phase, the RAG-1 and RAG-2 proteins (McBlane et al., 1995) recognise signal sequences and form a complex in which two minigenes with different lengths of signal sequences are placed directly next to each other. The RAG-1/RAG-2 proteins then introduce cuts into the genomic DNA at the boundary between the signal sequences and the minigenes through a reaction comprising several stages. The outcome of the process is the formation of two types of DNA ends: the blunt ends of the signal parts and the hairpin-like ends of the coding parts (Roth et al., 1992; Schlissel et al., 1993). In turn, DNA repair engages another group of proteins that are responsible for the process of non-homologous end joining (NHEJ). The Ku70/Ku80 protein heterodimer (Gu et al., 1997; Nussenzweig et al., 1996) interacts with the RAG-1/RAG-2 complex to bind to the coding ends. The catalytic subunit of DNA-dependent protein kinase (DNA-PK_{cs}) binds with Ku70/Ku80 and other proteins of the NHEJ pathway. DNA-PK_{cs} activates the Artemis endonuclease, which opens the hairpin structures located at the ends of the coding parts (Ma et al., 2002). Because the cuts are rarely symmetrical, the reaction yields DNA ends which contain single-strand segments that are either degraded by exonucleases or filled by polymerase through the addition of palindromic (P) nucleotides (Lafaille et al., 1989). Additional nucleotides (N) may be added to the coding sequences without the use of a matrix (i.e., without the use of a second strand of DNA as a model) (Alt & Baltimore, 1982) through a reaction catalysed by terminal deoxynucleotidyl transferase (TdT). At the end of the recombination process, the XRCC4-XLF proteins-DNA ligase 4 complex binds together the coding ends, thus restoring the continuity of genomic DNA (Ahnesorg et al., 2006). The signal ends are also joined together by the proteins of the NHEJ pathway to form a circular structure containing both signal sequences and the DNA segment that initially divided the minigenes which were bound during recombination.

To better understand the extraordinary efficiency of V(D)J recombination in generating the diversity of genes that encode antigen receptors, let us briefly analyse the case of genes coding TCRαβ. The α chain of this receptor is coded by the V and J minigenes. The two minigenes occur in the genome in ~45 and ~55 variants, respectively. Thus, about 2,475 (45 × 55) different TCRα genes can theoretically be created. The TCRβ chain is coded by the V, D and J minigenes occurring in ~50, 2 and 12 variants, respectively, which means that 1,200 different β chains can be created. Therefore, the combinatorial diversity alone allows for the creation of about 3×10^6 different TCRαβs (2,475 TCRα × 1,200 TCRβ), i.e. about 100 times more than all other genes put together. The addition of Ps and Ns to the coding ends is part of the mechanism of the aforementioned joint diversity, which, along with combinatorial diversity, allows for the creation of about 10^{15} different TCRαβs – 10,000 more than the number of stars in our galaxy.

3 Selection of the Receptor Repertoire: V(D)J Recombination Control

The method for obtaining a wide variety of receptors has a significant drawback: the same mechanism that allows for the random joining of minigenes and the addition of random nucleotides also creates the risk of producing nonfunctional receptors or, even worse, receptors that target the body's own antigens (autoantigens), leading to autoimmunity. To prevent the formation of cells with dysfunctional or autoaggressive receptors, the maturation of T and B cells uses selection mechanisms. Let us analyse the development of lymphocytes based on T cells of the αβ line (Figure 3.3). At the stage of DN2 and DN3 cells (with the phenotypes CD4-8-CD25+CD44+ and CD4-8-CD25+CD44-) during the development of Tαβ cells in the thymus, the locus coding TCRβ undergoes recombination: the $D_β$ and $J_β$ minigenes are joined together first, followed by the $V_β$ and $D_βJ_β$ minigenes (Cobb et al., 2006; Krangel, 2003). If the recombination is successful, the newly-created TCRβ chain is expressed and, along with the pre-Tα chain (with which it creates a pre-TCR molecule) (Groettrup et al., 1993), sends a signal that saves cells from death and causes their rapid proliferation. This allows for the selection of cells in which the gene coding TCRβ was rearranged correctly and encodes a functional protein, i.e. a protein that can bind with pre-Tα, can be exported onto the surface of the cell and is able to send a signal inside the cell. The signal from the pre-TCR also shuts down recombination, preventing the recombination of the minigenes of the other allele and ensuring that only one β chain of TCR is produced in the cell (Löffert et al., 1996). The recombination of the TCRα locus, during which the $V_α$ and $J_α$ minigenes are joined together, takes place in double-positive cells (CD4+CD8+). The TCRαβ molecule can already be created at this stage, and the cells where the molecule is expressed undergo selection to test for the potential usefulness or harmfulness of the receptor. During the selection, autoantigens are presented to lymphocytes in the context of an MHC molecule, i.e., the same process is as it is later when non-self antigens are recognised during the immune system's normal response. A lack of interaction between TCR and the MHC/autoantigen complex blocks the death-preventing signal from entering the lymphocyte, which dies as a result (Kisielow, Teh, et al., 1988). Thus, lymphocytes with a useless receptor, i.e. one that is unable to recognise any antigen in the context of the MHC molecule or transmit the signal, are eliminated. A very strong signal resulting from a strong interaction between the lymphocyte and the presenting cell also leads to the death of the lymphocyte as it indicates that the receptor has recognised the autoantigen.[1] About 2% of lymphocytes survive both types of selection (positive and negative), i.e. those that, following

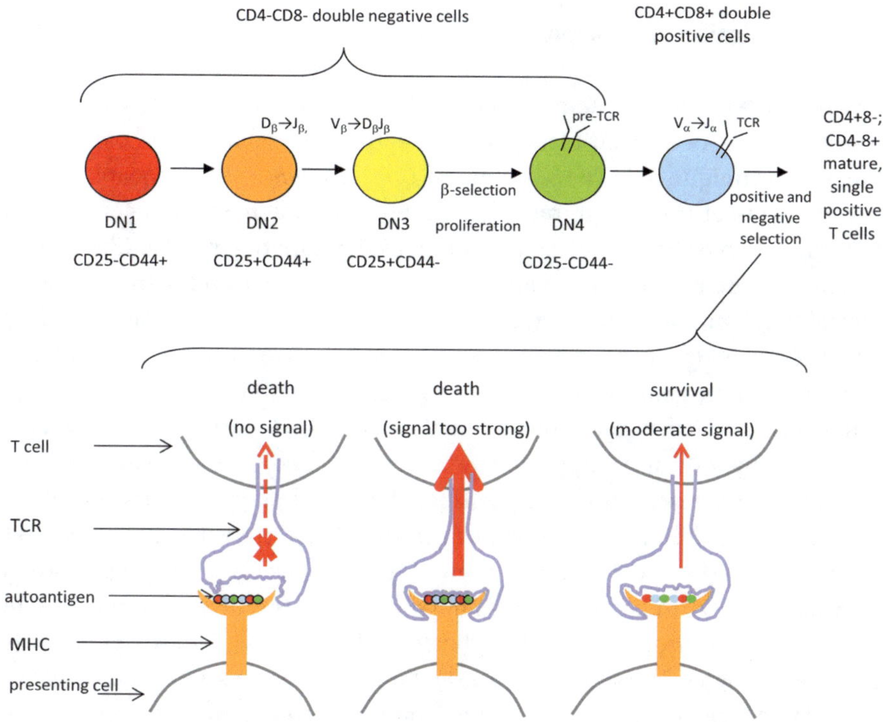

FIGURE 3.3 Maturation of Tαβ cells in the thymus and the selection of the TCR repertoire

interaction with the antigen-presenting cell, produce a moderate signal which is indicative of their ability to recognise antigens in the context of the MHC molecule, but not indicative of autoantigen recognition (Kisielow, Blüthmann, et al., 1988).

Another issue with V(D)J recombination is the potential harm of the process itself. Compromised DNA integrity during recombination, which involves cutting and combining DNA, carries the risk of errors that can lead to DNA translocation and neoplastic transformations. Thus, V(D)J recombination is controlled by many factors: the availability of DNA for *RAG-1*/*RAG-2* recombinase, tissue specificity, timing, the number of cells involved, and, most importantly, precise control over the expression and activity of *RAG-1* and *RAG-2*. The final section of this paper will discuss the evolutionary origin of *RAG-1* and *RAG-2* genes and the hypotheses explaining the mechanisms controlling their expression.

4 Evolution of Transposase

The adaptive immune system that takes advantage of T and B cells, which show a vast diversity of antigen-recognising receptors created during V(D)J recombination, is unique to jawed vertebrates (*Gnathostomata*). This means that all key elements of the system must have evolved between 450 and 500 million years ago (between the divergence period of the Agnatha and the Chondrichthyes). Researchers who have attempted to find the reason for the immunological "big bang", as it is referred to, have proposed the hypothesis that *RAG-1* and *RAG-2*, i.e., the genes which code the enzymes required for recombination, entered the genome of a common ancestor of jawed vertebrates through a horizontal transfer as a mobile genetic element, i.e. a transposon (Schatz, 2004). The hypothesis is supported by the fact that in biochemical terms, V(D)J recombination resembles transposition, the signal sequences (which flank the minigenes) resemble the sequences that flank transposons (Cowell et al., 2004) and the RAG-1 and RAG-2 proteins themselves are able to perform a transposition, i.e. to cut away a DNA fragment and insert it somewhere else in the genome (*in vitro*, but also, though very rarely, *in vivo*) (Agrawal et al., 1998; Hiom et al., 1998; Messier et al., 2003). Furthermore, *RAG-1* and *RAG-2* genes are located directly next to each other, and their protein-coding segments are enclosed within single exons (Oettinger et al., 1990). This compact structure resembles that of transposases (enzymes coded by transposons). The hypothesis concerning the transposon origin of RAG recombinase has been reinforced by the discovery of transposases from the Transib family in many species of invertebrates whose structures resemble fragments of the RAG-1 protein (Kapitonov & Jurka, 2005). Thus, the evolution of the whole system would be as follows (Figure 3.4): a transposon coding RAG transposase was integrated with the genome of a common ancestor of jawed vertebrates, and a DNA fragment that contained only signal sequences and whose mobility likely still depended on the enzymatic activity of transposase was created.[2] The transposition of such a fragment into the coding part of the pre-receptor gene would make the expression of a functional version of the receptor fully dependent on the activity of RAG transposase, which in turn would be the only enzyme able to restore the continuity of the open reading frame. A series of duplications and mutations of the pre-receptor gene would lead to the creation of the minigenes and all loci coding the antigen receptors known today. The integration of the *RAG* transposon, in addition to the potential benefit of diversifying antigen recognition receptors, must have enforced the evolution of mechanisms for preventing threats to genome stability arising from the activity of transposase.

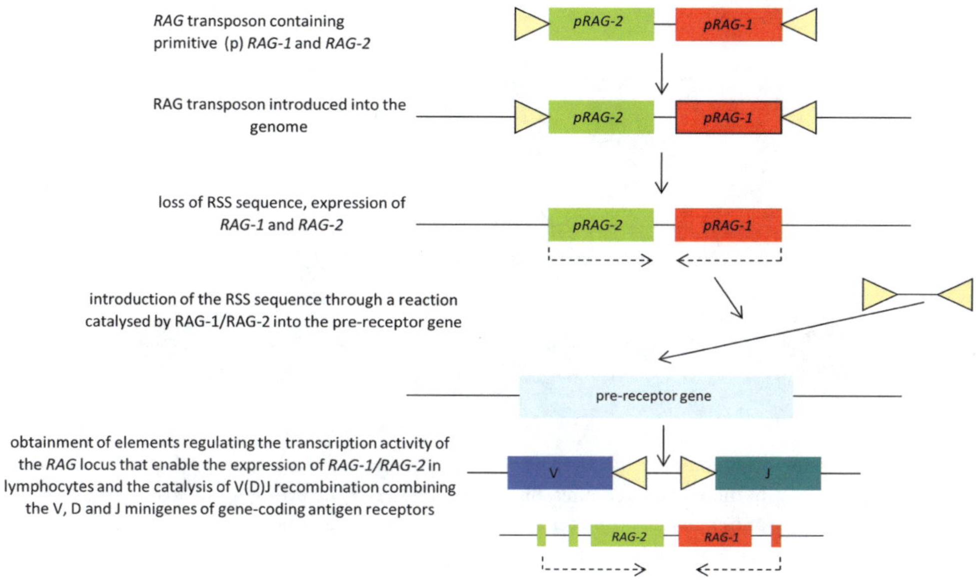

FIGURE 3.4 The scenario of the transposon hypothesis for the origin of *RAG-1* and *RAG-2*

These mechanisms would include changes to the biochemical properties of the RAG-1/RAG-2 complex that would limit its ability to catalyse transposition to a minimum while maintaining its ability to cleave DNA. These changes would also enable interaction with the proteins of the DNA repair system and the evolution of mechanisms for limiting the expression of RAG recombinase to particular types of cells.

Let us, however, go back to the very beginning when the *RAG* transposon was introduced into the genome of a common ancestor of jawed vertebrates. One should bear in mind that genomes are quite literally "littered" with the remains of many transposons that, throughout evolution, "infected" the genetic material of their hosts. The integration of transposons with the host genome is (from the host's point of view) harmful, primarily due to the danger caused by the uncontrolled ability of the transposon to move within the genome, potentially damaging the host's genes. Consequently, organisms have evolved defense mechanisms that help hamper the activity of transposons, mainly by silencing their expression through epigenetic changes in DNA structure (DNA methylation and/or histone modifications). How, then, can the fact that the transposon containing the *RAG-1* and *RAG-2* pre-genes was not silenced be explained? Analysis of the few cases where genes originating from transposons

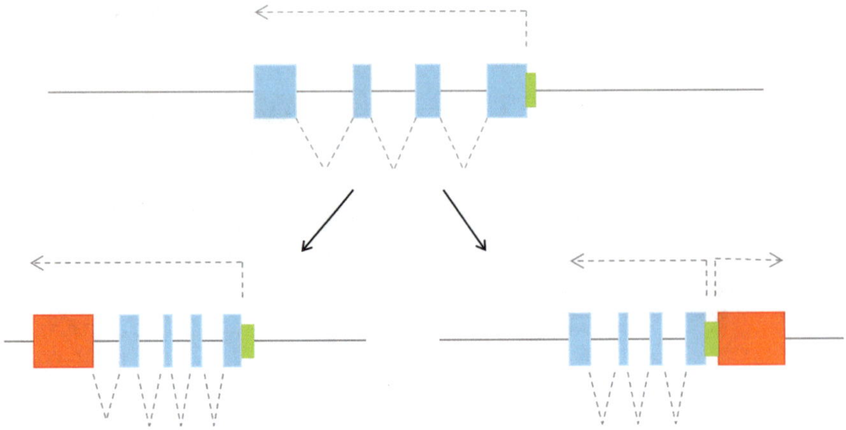

FIGURE 3.5 Characteristic locations of genes originating from transposons that still function in their hosts' genomes. Blue rectangles mark the exons of the host's gene, green ones mark its promoter and red ones mark transposase genes. Arrows indicate the direction of transcription

are still active showed that these transposons have integrated with the genome in very specific locations (Kalitsis & Saffery, 2009; Figure 3.5).

One such location is a DNA fragment located directly downstream of an existing gene. In such a case, later changes create a hybrid gene coding a protein with a modified function (compared to that of transposase and the original host protein). Another characteristic location is the head-to-head arrangement, whereby the transposon integrates itself above an existing gene and the directions of the host gene and transposase expression are opposite (Kalitsis & Saffery, 2009). It has been shown that in such cases, the host's genes have a constitutionally active two-directional promoter that may trigger the expression of both the host's gene and the sequences located above it. From the viewpoint of the transposon, the second option is more beneficial, as after integrating with the genome, the transposase gene will have a guaranteed mechanism of expression and any attempts at silencing it would also hamper the expression of the host's gene. Furthermore, the structure of the host's gene and the transposase gene is left intact (in contrast to the first option), which allows both genes to perform their original functions. There are very good reasons to conclude that the integration of the *RAG* transposon involved the second of the aforementioned mechanisms and the properties of the host's *NWC* gene enabled the integration and survival of the *RAG* transposon.

The *NWC* gene (Cebrat et al., 2005) is the third gene in the locus containing *RAG-1* and *RAG-2*. The location of the locus is conserved in all species of jawed vertebrates (Figure 3.6). The orthologs of the *NWC* gene are also present

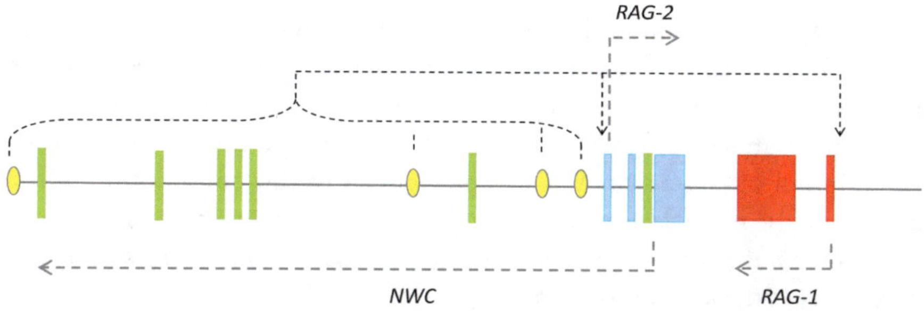

FIGURE 3.6 Structure of the *RAG/NWC* locus. The green rectangles mark the exons of the *NWC* genes, the blue rectangles mark *RAG-2* and the red rectangles mark *RAG-1*. Horizontal arrows show the directions and range of transcription. Vertical arrows show the effect of *cis*-regulatory elements (yellow ovals) on the promoters of *RAG-1* and *RAG-2*

in many invertebrate species, which indicates that the *NWC* locus may have been where the integration of the *RAG* transposon took place. Further indication supporting this hypothesis is the fact that the coding part of *RAG-2* gene is located immediately upstream of the *NWC* gene (Cebrat et al., 2005), *RAG-2* and *NWC* genes undergo convergent transcription, the promoter of the *NWC* gene shows constitutional activity (Cebrat et al., 2008) and, most importantly, the *NWC* promoter shows bidirectional activity (Laszkiewicz et al., 2012).

As it has been stated in previous sections of this chapter, the expression of *RAG-1* and *RAG-2* falls under very precise control: it is well-coordinated and takes part in specific moments of T and B cell development. The elements responsible for this control are *cis*-regulatory elements. These are sequences located predominantly on the 5' side of the RAG-2 gene which, once the appropriate transcription factors are attached, interact with *RAG-1* and *RAG-2* promoters, leading to their activation in the maturing lymphocytes (Figure 3.6) (Hsu et al., 2003; Wei et al., 2002; Yannoutsos et al., 2004; Yu et al., 1999). The control elements for the expression of *RAG*s (both the *cis*-regulatory elements and the promoters) are considered a later evolutionary acquisition. The promoter of the *NWC* gene, in contrast to the *RAG*s, is active in non-lymphoid cells (Cebrat et al., 2005, 2008). The constitutive activity of the *NWC* promoter originated from its other properties which, as mentioned above, were found to have been key during the first stages of coexistence between the *RAG* transposon and the genome of the common ancestor of jawed vertebrates. However, this activity could have been harmful to the correct functioning of the control mechanisms for the expression of *RAG*s during lymphocyte maturation. This may stem from the fact that the transcription of the *NWC* gene, involving the area with *cis*-regulatory elements that activate *RAG* promoters, could interfere with the binding of transcriptional factors to these sequences in the process known as transcriptional interference (Shearwin et al., 2005). However, such interference does not occur thanks to the silencing of the *NWC* gene in lymphocytes (Cebrat et al., 2008) by means of lymphocyte-specific transcription repressors and the transcription of *RAG-2*, which is complementary to the transcription of *NWC* and causes the methylation of the *NWC* promoter (Laszkiewicz et al., 2014).

It is fascinating that the activity of what in essence constitutes a molecular pathogen, i.e., the *RAG* transposon, ultimately led to the development of a seemingly simple, yet complex system that is very effective at combating other pathogens and determines the body's molecular "self-awareness". However, one should remember that the evolution of this system also depended on the interaction between the *RAG* transposon and the host's genome. A complete characterisation of the mutual relationships in the control mechanisms for the expression of *RAG* and *NWC* genes, an investigation into the control mechanisms for the expression of *NWC* genes in invertebrates, and the characterization of the NWC protein function should improve our future understanding of the evolutionary processes that have led to the emergence of the adaptive immune system.

Notes

1 Some of the lymphocytes that survive the strong signal resulting from the recognition of the autoantigen constitute the exceptional CD4+8- subpopulation of T cells, called regulatory lymphocytes (Wojciech et al., 2014), which are later used to control the action of conventional lymphocytes, e.g. when the immune response is silenced, and which constitute the second line of defence (after the selection in the thymus) against autoaggresive lymphocytes.

2 Perhaps this occurred as a result of the duplication of the *RAG* transposon and the incomplete transposition of the duplicate into another location in the genome.

References

Agrawal, A., Eastman, Q. M., & Schatz, D. G. (1998). Transposition mediated by RAG1 and RAG2 and its implications for the evolution of the immune system. *Nature*, *394*(6695), 744–751.

Ahnesorg, P., Smith, P., & Jackson, S. P. (2006). XLF interacts with the XRCC4-DNA ligase IV complex to promote DNA nonhomologous end-joining. *Cell*, *124*(2), 301–313.

Alt, F. W., & Baltimore, D. (1982). Joining of immunoglobulin heavy chain gene segments: implications from a chromosome with evidence of three D-JH fusions. *Proceedings of the National Academy of Sciences of the United States of America*, *79*(13), 4118–4122.

Bassing, C. H., Swat, W., & Alt, F. W. (2002). The mechanism and regulation of chromosomal V(D)J recombination. *Cell, 109*(2), S45–S55.

Bengtén, E., Wilson, M., Miller, N., Clem, L. W., Pilström, L., & Warr, G. W. (2000). Immunoglobulin isotypes: Structure, function, and genetics. *Current Topics in Microbiology and Immunology, 248*, 189–219.

Cebrat, M., Cebula, A., Laszkiewicz, A., Kasztura, M., Miazek, A., & Kisielow, P. (2008). Mechanism of lymphocyte-specific inactivation of RAG-2 intragenic promoter of NWC: Implications for epigenetic control of RAG locus. *Molecular Immunology, 45*(8), 2297–2306.

Cebrat, M., Miazek, A., & Kisielow, P. (2005). Identification of a third evolutionarily conserved gene within the RAG locus and its RAG1-dependent and -independent regulation. *European Journal of Immunology.* https://onlinelibrary.wiley.com/doi/abs/10.1002/eji.200526225

Cobb, R. M., Oestreich, K. J., Osipovich, O. A., & Oltz, E. M. (2006). Accessibility control of V(D)J recombination. *Advances in Immunology, 91*, 45–109.

Cowell, L. G., Davila, M., Ramsden, D., & Kelsoe, G. (2004). Computational tools for understanding sequence variability in recombination signals. *Immunological Reviews, 200*, 57–69.

Eastman, Q. M., Leu, T. M. J., & Schatz, D. G. (1996). Initiation of V(D)J recombination in vitro obeying the 12/23 rule. *Nature, 380*(6569), 85–88.

Groettrup, M., Ungewiss, K., Azogui, O., Palacios, R., Owen, M. J., Hayday, A. C., & von Boehmer, H. (1993). A novel disulfide-linked heterodimer on pre-T cells consists of the T cell receptor β chain and a 33 kd glycoprotein. *Cell, 75*(2), 283–294.

Gu, Y., Seidl, K. J., Rathbun, G. A., Zhu, C., Manis, J. P., van der Stoep, N., Davidson, L., Cheng, H. L., Sekiguchi, J. M., Frank, K., Stanhope-Baker, P., Schlissel, M. S., Roth, D. B., & Alt, F. W. (1997). Growth retardation and leaky SCID phenotype of Ku70-deficient mice. *Immunity, 7*(5), 653–665.

Hiom, K., Melek, M., & Gellert, M. (1998). DNA transposition by the RAG1 and RAG2 proteins: A possible source of oncogenic translocations. *Cell, 94*(4), 463–470.

Hsu, L.-Y., Lauring, J., Liang, H.-E., Greenbaum, S., Cado, D., Zhuang, Y., & Schlissel, M. S. (2003). A conserved transcriptional enhancer regulates RAG gene expression in developing B cells. *Immunity, 19*(1), 105–117.

Kalitsis, P., & Saffery, R. (2009). Inherent promoter bidirectionality facilitates maintenance of sequence integrity and transcription of parasitic DNA in mammalian genomes. *BMC Genomics, 10*, 498.

Kapitonov, V. V., & Jurka, J. (2005). RAG1 core and V(D)J recombination signal sequences were derived from Transib transposons. *PLoS Biology, 3*(6), e181.

Kisielow, P., Blüthmann, H., Staerz, U. D., Steinmetz, M., & von Boehmer, H. (1988). Tolerance in T-cell-receptor transgenic mice involves deletion of nonmature CD4+8+ thymocytes. *Nature, 333*(6175), 742–746.

Kisielow, P., Teh, H. S., Blüthmann, H., & von Boehmer, H. (1988). Positive selection of antigen-specific T cells in thymus by restricting MHC molecules. *Nature, 335*(6192), 730–733.

Krangel, M. S. (2003). Gene segment selection in V(D)J recombination: Accessibility and beyond. *Nature Immunology, 4*(7), 624–630.

Lafaille, J. J., DeCloux, A., Bonneville, M., Takagaki, Y., & Tonegawa, S. (1989). Junctional sequences of T cell receptor gamma delta genes: Implications for gamma delta T cell lineages and for a novel intermediate of V-(D)-J joining. *Cell, 59*(5), 859–870.

Laszkiewicz, A., Bzdzion, Ł., Kasztura, M., Snieżewski, L., Janik, S., Kisielow, P., & Cebrat, M. (2014). Ikaros and RAG-2-mediated antisense transcription are responsible for lymphocyte-specific inactivation of NWC promoter. *PLoS One, 9*(9), e106927.

Laszkiewicz, A., Sniezewski, L., Kasztura, M., Bzdzion, L., Cebrat, M., & Kisielow, P. (2012). Bidirectional activity of the NWC promoter is responsible for RAG-2 transcription in non-lymphoid cells. *PLoS One, 7*(9), e44807.

Lewis, S. M. (1994). The mechanism of V(D)J joining: Lessons from molecular, immunological, and comparative analyses. *Advances in Immunology, 56*, 27–150.

Litman, G. W., Anderson, M. K., & Rast, J. P. (1999). Evolution of antigen binding receptors. *Annual Review of Immunology, 17*, 109–147.

Löffert, D., Ehlich, A., Müller, W., & Rajewsky, K. (1996). Surrogate light chain expression is required to establish immunoglobulin heavy chain allelic exclusion during early B cell development. *Immunity, 4*(2), 133–144.

Ma, Y., Pannicke, U., Schwarz, K., & Lieber, M. R. (2002). Hairpin opening and overhang processing by an Artemis/DNA-dependent protein kinase complex in nonhomologous end joining and V(D)J recombination. *Cell, 108*(6), 781–794.

McBlane, J. F., van Gent, D. C., Ramsden, D. A., Romeo, C., Cuomo, C. A., Gellert, M., & Oettinger, M. A. (1995). Cleavage at a V(D)J recombination signal requires only RAG1 and RAG2 proteins and occurs in two steps. *Cell, 83*(3), 387–395.

Messier, T. L., O'Neill, J. P., Hou, S., Nicklas, J. A., & Finette, B. A. (2003). *In vivo* transposition mediated by V(D)J recombinase in human T lymphocytes. *The EMBO Journal, 22*(6), 1381–1388.

Nussenzweig, A., Chen, C., da Costa Soares, V., Sanchez, M., Sokol, K., Nussenzweig, M. C., & Li, G. C. (1996). Requirement for Ku80 in growth and immunoglobulin V(D)J recombination. *Nature, 382*(6591), 551–555.

Oettinger, M. A., Schatz, D. G., Gorka, C., & Baltimore, D. (1990). RAG-1 and RAG-2, adjacent genes that synergistically activate V(D)J recombination. *Science, 248*(4962), 1517–1523.

Ramsden, D. A., Baetz, K., & Wu, G. E. (1994). Conservation of sequence in recombination signal sequence spacers. *Nucleic Acids Research, 22*(10), 1785–1796.

Roth, D. B., Menetski, J. P., Nakajima, P. B., Bosma, M. J., & Gellert, M. (1992). V(D)J recombination: broken DNA molecules with covalently sealed (hairpin) coding ends in scid mouse thymocytes. *Cell, 70*(6), 983–991.

Sakano, H., Hüppi, K., Heinrich, G., & Tonegawa, S. (1979). Sequences at the somatic recombination sites of immunoglobulin light-chain genes. *Nature, 280*(5720), 288–294.

Schatz, D. G. (2004). Antigen receptor genes and the evolution of a recombinase. *Seminars in Immunology, 16*(4), 245–256.

Schlissel, M., Constantinescu, A., Morrow, T., Baxter, M., & Peng, A. (1993). Double-strand signal sequence breaks in V(D)J recombination are blunt, 5'-phosphorylated, RAG-dependent, and cell cycle regulated. *Genes & Development, 7*(12B), 2520–2532.

Shearwin, K. E., Callen, B. P., & Egan, J. B. (2005). Transcriptional interference–a crash course. *Trends in Genetics: TIG, 21*(6), 339–345.

Tonegawa, S. (1983). Somatic generation of antibody diversity. *Nature, 302*(5909), 575–581.

Wei, X.-C., Kishi, H., Jin, Z.-X., Zhao, W.-P., Kondo, S., Matsuda, T., Saito, S., & Muraguchi, A. (2002). Characterization of chromatin structure and enhancer elements for murine recombination activating gene-2. *Journal of Immunology, 169*(2), 873–881.

Wojciech, L., Ignatowicz, A., Seweryn, M., Rempala, G., Singh Pabla, S., McIndoe, R. A., Kisielow, P., & Ignatowicz, L. (2014). The same self-peptide selects conventional and regulatory CD4+ T cells with identical antigen receptors. *Nature Communications, 5*(1), 1–10.

Yannoutsos, N., Barreto, V., Misulovin, Z., Gazumyan, A., Yu, W., Rajewsky, N., Peixoto, B. R., Eisenreich, T., & Nussenzweig, M. C. (2004). A cis element in the recombination activating gene locus regulates gene expression by counteracting a distant silencer. *Nature Immunology, 5*(4), 443–450.

Yu, W., Misulovin, Z., Suh, H., Hardy, R. R., Jankovic, M., Yannoutsos, N., & Nussenzweig, M. C. (1999). Coordinate regulation of RAG1 and RAG2 by cell type-specific DNA elements 5' of RAG2. *Science, 285*(5430), 1080–1084.

CHAPTER 4

Degrees of Complexity in the Recognition of Substrates by Proteolytic Enzymes

Marcin Drąg

Abstract

Proteases constitute a group of approximately 700 enzymes in humans, and are responsible for the regulation of almost all metabolic pathways. Disturbance of their activity leads to serious diseases, the so-called "diseases of civilization", including diabetes, cancers and neurodegenerative disorders. A defining feature of proteases is their ability to bind protein substrates. Exopeptidases exhibit a comparatively simple substrate-binding mechanism, whereas endoproteases employ a complex process for substrate recognition and binding. This chapter highlights characteristic examples of both types of proteases illustrating the spectrum of mechanisms employed by these enzymes.

Keywords

protease – substrate specificity – combinatorial chemistry – unnatural amino acids – natural amino acids

1 Introduction

Proteolytic enzymes, i.e., proteases, are among the most important enzymes active in the processes of monitoring metabolic pathways in living organisms (Drag & Salvesen, 2010; Turk, 2006). Hydrolysis of the peptide binding of the protein and peptide substrates catalyzed by them is considered the key irreversible post-translational modification. Substrate hydrolysis by proteases is strictly controlled, and any imbalance in this system leads to disorders in their functioning (Drag & Salvesen, 2010). An apt example is the process of controlled cell death, known as apoptosis. It is strictly controlled by proteases from the group of caspases (G. S. Salvesen & Dixit, 1997). Disorders of the activity of caspases may lead to two types of diseases depending on the

DOI:10.1163/9789004714540_004

grade of their activity. Excessive apoptosis leads to uncontrolled cellular atrophy, which fundamentally influences the development of neurodegenerative diseases. Conversely, insufficient apoptosis eliminating damaged cells leads to their uncontrolled proliferation, which causes cancer development (Pop & Salvesen, 2009).

Proteases are classified according to their mechanism of action and the location of hydrolysis (Drag & Salvesen, 2010). In the latter classification, proteases from the group of exopeptidases hydrolyze amino acids from the N-terminus (aminopeptidases) or C-terminus (carboxypeptidases) of the substrate, while a very large group of proteases hydrolyzing substrates in the middle of the peptide chain constitute the group of endopeptidases. Recognition of substrates is based on the specific fit of the twenty amino acid residues encoded by DNA into the binding pocket of the protease. These amino acids can be bound to the C-terminus of the protein substrate, counting from the site of hydrolysis, and then depending on the pocket they are referred to as P1′, P2′ and Pn′. Amino acids that are bound to the N-terminus of the protein substrate are termed P1, P2 and Pn (Deu et al., 2012; M. Poreba & Drag, 2010).

2 From Simplicity to Complexity of Proteases and Their Binding Mechanisms

Depending on the type of protease, the mechanism of binding substrates can be relatively simple or complex. Therefore, for example, the simplest system can be distinguished in the case of the proteolytic enzymes from the group of aminopeptidases or carboxypeptidases, which hydrolyze single amino acids from the ends of the peptide chain (Drag & Salvesen, 2010). For instance, in the case of aminopeptidases, which hydrolyze amino acids from the N-terminus of the peptide, the key is to recognize first of all the amino acid binding in the S1 pocket (i.e., enzyme's subsite/binding site), because only this one is hydrolyzed. Of course, in the aspect of the mechanism of binding of the whole substrate, a certain role may also be played by the S1′ or S2′ pocket, but it is considered that the determinant of the specificity is precisely the binding of the amino acid to the S1 pocket. This is confirmed by numerous studies on the substrate specificity of aminopeptidases applying fluorogenic substrate libraries consisting of natural as well as unnatural (not coded by DNA) amino acids and reporter groups in the form of a fluorophore (Kasperkiewicz et al., 2012). Moreover, considering the preferences of aminopeptidases, we can carry out further analysis and distinguish those proteases that are characterized by high simplicity in the aspect of substrate specificity from those that are able to recognize a complex palette of amino acids of varied structural characteristics.

The simplest proteases undoubtedly include methionine aminopeptidases. An analysis of the substrate specificity of these enzymes showed that among all natural amino acids, they practically exclusively recognize methionine, while among the unnatural ones, only a few are structural analogs of methionine (Marcin Poreba, Gajda, et al., 2012). This is not particularly surprising, as the role of methionine aminopeptidases is the hydrolysis of N-terminal methionine with freshly expressed proteins in ribosomes. A considerably more complex system can be observed in the case of such aminopeptidases as leukotriene aminopeptidase (LT4AH), aminopeptidase N (APN) or for example leucyl aminopeptidase (Byzia et al., 2014; Drag et al., 2010; Marcin Poreba, McGowan, et al., 2012; Węglarz-Tomczak et al., 2013). These enzymes demonstrate great tolerance for natural amino acids and unnatural amino acids. For example, leukotriene aminopeptidase shows several dozen times greater affinity towards selected unnatural amino acids (e.g., benzyl esters of aspartic acid and glutamic acid, homoarginine) compared to its best natural amino acid, which is arginine (Byzia et al., 2014). The increased specificity and simultaneous complexity of recognized structures result from the function of these enzymes that are not as specialized as methionine aminopeptidases. All aminopeptidases with broad substrate specificity are responsible for processing numerous types of protein substrates and peptides, necessitating their ability to recognize many types of structural amino acids.

A somewhat more complex system is observed in the case of dipeptidyl aminopeptidases, which hydrolyze the dipeptide fragment from the substrate N-terminus. A classic example is cathepsin C, a lysosomal cysteine protease belonging to the papain protease family that participates in the activation of numerous other proteases, and is also involved in the development of diseases such as Haim-Munk syndrome and Papillon-Lefèvre syndrome (Mohamed & Sloane, 2006; Yuan et al., 2006). The multitasking ability of this enzyme implicates almost automatically its broad substrate specificity, and in consequence structural complexity of recognized amino acids (Wang et al., 2011). Moreover, this enzyme is also more complex in terms of the number of amino acids that it binds to and recognizes in order to function correctly. Therefore, cathepsin C has very well-defined pockets binding S1 and S2, and also the pockets S1′ and S2′. Studies on substrate specificity using a targeted fluorogenic substrate library towards the S1 and S2 pockets applying human, bovine and malaria orthologs showed that in the S1 pocket, mammalian orthologs present a much more complex palette of recognized natural and unnatural amino acids. The most active amino acid in the case of all three enzymes was L-Nle(O-Bzl), while

L-Bpa was only recognized by mammalian orthologs and not by malaria. Then in the S2 pocket, great differentiation between the human and bovine cathepsin C and its malaria ortholog was visible. Mammalian orthologs preferred in this position L-Abu, which is an alanine analog longer by one methylene group, while malaria cathepsin C definitely preferred in this position L-Pip, which is a proline analog longer by one methylene group. L-Pip also stood out as being very well recognizable by malaria enzymes but very poorly recognizable by mammalian orthologs of cathepsin C (Marcin Poreba et al., 2014).

Certainly, the most complex system for recognizing specific amino acids in peptide and protein substrates is that of proteolytic enzymes from the group of endopeptidases. These enzymes constitute the largest group of proteases and are involved in key processes controlling metabolic pathways. One of the more important groups within this category is caspases, which are involved in apoptosis (Pop & Salvesen, 2009).It is a very complex system of enzymes that are divided into initiator caspases (caspase 2, 8, 9 and 10) and executioner caspases (caspase 3, 6 and 7). For the apoptosis to proceed correctly, initiator caspases have to be activated by activating factors, and then active initiator caspases activate executioner procaspases into active enzymes (Riedl & Salvesen, 2007; Guy S. Salvesen & Riedl, 2008).

The whole process involves activation by hydrolysis of executioner caspases, which subsequently hydrolyze protein substrates inside the cell. A key aspect is the ability of individual caspases to recognize specific peptide sequences. Notably, all caspases have the specificity to exclusively recognize aspartic acid in position P1, while positions P2-P4 are somewhat more complex. Practically all caspases recognize the most natural amino acids in these positions. This confirms the broad range of substrates processed by them. Interestingly, many caspases have very similar profiles of substrate specificity. This was shown by the studies on the substrate specificity of all caspases using a tetrapeptide combinatorial library of fluorogenic substrates (Thornberry et al., 1997). Examples of such caspases include caspases 4 and 5, caspases 3 and 7, and caspases 6 and 8. This interesting example shows that even a complicated palette of 20 natural amino acids is not sufficient to differentiate similar enzymes in the three binding pockets (S2–S4). Interestingly, many of these sequences were used in key biological studies concerning caspases, with researchers carrying out these studies often unaware that the substrates they used were recognized by more than one enzyme. The complexity of an apoptotic system caused several caspases to be active simultaneously, and the activity registered using fluorogenic or chromogenic substrates reflects the combined activity of these caspases. A few years ago, the groups of Salvesen and Green showed for the first time that this is a major problem while interpreting results, and proved at the same time that in the direct comparative analysis, substrates based on these sequences cannot be used for individual caspases (McStay et al., 2008). These studies were independently confirmed by the group of Bogyo, who used inhibitors and small molecule chemical markers based on these sequences (Berger et al., 2006). Thus, it was unquestionably demonstrated that the system consisting of 20 amino acids is too simple to recognize similar endoproteases. This problem was solved to a large extent by using a more complex combination of amino acids in the studies, that is natural amino acids and a very large group (about 100) of unnatural amino acids (M. Poreba et al., 2014). The created tetrapeptide hybrid combinatorial library of fluorogenic substrates was used to profile the substrate specificity of all apoptotic caspases. The studies showed that using unnatural amino acids increases the chance to find many more specific sequences of amino acids than those, which exclusively consist of natural amino acids. The grade of specificity of some sequences was so great that they were successfully used to clarify the processes of procaspase 9 activation to caspase 9, and then to the activation of procaspase 3. This example clearly shows that some apparently complex systems have a simple system, and much more complicated and advanced solutions are necessary for their differentiation.

The next interesting example is that of neutrophil serine proteases, which include neutrophil elastase, cathepsin G, proteinase 3 and NSP4 (Perera et al., 2013; Pham, 2006). These enzymes occur in neutrophils and are responsible for their correct functioning. However, the disturbance of their activity, for instance, in the case of neutrophil elastase may lead to the development of lung cancer and other lung diseases (Moroy et al., 2012). Moreover, these enzymes are involved in the process of creating active neutrophil extracellular traps responsible for combating pathogens (Brinkmann et al., 2004; O'Donoghue et al., 2013). These proteases were very well characterized and described. The substrate specificity towards natural amino acids of all these proteases is also known (Perera et al., 2013; Rawlings et al., 2014; Schilling & Overall, 2008). However, the greatest problem was creating specific chemical tools for their study. Similarly, as in the case of caspases, these enzymes recognized similar sequences, which prevented effective studies of their functions in the biological system. Particularly difficult here were neutrophil elastase and proteinase 3, which showed a very similar profile of substrate specificity (Kasperkiewicz et al., 2014). What is more, the existing substrates and inhibitors based on simple natural amino acids were not sufficiently active in biological systems.To address this issue the solution was to replace simple substrate libraries with much more complex hybrid fluorogenic substrate libraries containing natural amino acids as well as an extensive collection of unnatural amino acids. The performed studies on substrate specificity allowed the selection of a substrate that was several hundred times more selective towards neutrophil elastase than proteinase 3, and, what is more, this substrate was several thousand times more active compared to the commercially available sequence consisting of natural amino acids. These properties also enabled the creation of a small molecule chemical marker with the help of which it was possible to visualize without difficulty the activity of human neutrophil elastase in the process of neutrophil extracellular traps. It was also shown here that it is possible to obtain substrates with a relatively simple structure (compared to the protein substrates) and at the same time complex enough to obtain a desired grade of specificity and activity.

Another interesting example of such simplicity and simultaneous complexity is that comprising the whole group of deubiquitinating endoproteases (DUB proteases), whose function is the activation and processing of ubiquitin (Russell & Wilkinson, 2005). The complexity of this system is such that to date over a hundred enzymes belonging to this family have successfully been characterized and identified. Its simplicity, on the other hand, consists in the fact that practically all these enzymes recognize only one substrate, the small protein ubiquitin. This phenomenon is fascinating in that many of these enzymes act at the same time and in the same place, and it is very difficult to understand their co-dependence in the processing of ubiquitin (Eletr & Wilkinson, 2014). What is more, the latest studies prove that many of these enzymes specialize in processing specific chain forms of polyubiquitin (Reyes-Turcu et al., 2009). Analogical to this are desumoylating proteases (SENPs) which process SUMO protein substrates structurally similar to ubiquitin (SUMO-1, 2 and 3) (Mikolajczyk et al., 2007). Also in this case it is not fully known at present which mechanisms direct this whole system. It is certainly known that deubiquitinating proteases as well as desumoylating proteases have substrate specificity determined with the help of combinatorial fluorogenic substrate libraries in the S1-S4 pockets very similar to those occurring in their natural substrates, that is LRGG for ubiquitin and QTGG for SUMO (Drag, Mikolajczyk, Bekes, et al., 2008; Drag, Mikolajczyk, Krishnakumar, et al., 2008). One of the greatest challenges at present is to investigate these systems, particularly taking into consideration the great medical value of these enzymes which participate in the development processes of cancers and neurodegenerative diseases (Lim & Baek, 2013).

3 Conclusions

In summary, it can be concluded that proteolytic enzymes, regardless of their classification based on the site of hydrolysis of peptide bonds, can have either straightforward substrate specificity or, what is more often observed, a highly complex mechanism for substrate recognition. Research in this field is currently advanced, yet hundreds of proteases still remain to be studied and systematized, which makes this research subject unusually fascinating.

References

Berger, A. B., Sexton, K. B., & Bogyo, M. (2006). Commonly used caspase inhibitors designed based on substrate specificity profiles lack selectivity. *Cell Research, 16*(12), 961–963.

Brinkmann, V., Reichard, U., Goosmann, C., Fauler, B., Uhlemann, Y., Weiss, D. S., Weinrauch, Y., & Zychlinsky, A. (2004). Neutrophil extracellular traps kill bacteria. *Science, 303*(5663), 1532–1535.

Byzia, A., Haeggström, J. Z., Salvesen, G. S., & Drag, M. (2014). A remarkable activity of human Leukotriene A4 Hydrolase (LTA4H) toward unnatural amino acids. *Amino Acids, 46*(5), 1313–1320.

Deu, E., Verdoes, M., & Bogyo, M. (2012). New approaches for dissecting protease functions to improve probe development and drug discovery. *Nature Structural & Molecular Biology, 19*(1), 9–16.

Drag, M., Bogyo, M., Ellman, J. A., & Salvesen, G. S. (2010). Aminopeptidase fingerprints, an integrated approach for identification of good substrates and optimal inhibitors. *The Journal of Biological Chemistry, 285*(5), 3310–3318.

Drag, M., Mikolajczyk, J., Bekes, M., Reyes-Turcu, F. E., Ellman, J. A., Wilkinson, K. D., & Salvesen, G. S. (2008). Positional-scanning fluorigenic substrate libraries reveal unexpected specificity determinants of DUBs (deubiquitinating enzymes). *Biochemical Journal, 415*(3), 367–375.

Drag, M., Mikolajczyk, J., Krishnakumar, I. M., Huang, Z., & Salvesen, G. S. (2008). Activity profiling of human deSUMOylating enzymes (SENPs) with synthetic substrates suggests an unexpected specificity of two newly characterized members of the family. *Biochemical Journal, 409*(2), 461–469.

Drag, M., & Salvesen, G. S. (2010). Emerging principles in protease-based drug discovery. *Nature Reviews. Drug Discovery, 9*(9), 690–701.

Eletr, Z. M., & Wilkinson, K. D. (2014). Regulation of proteolysis by human deubiquitinating enzymes. *Biochimica et Biophysica Acta, 1843*(1), 114–128.

Kasperkiewicz, P., Gajda, A. D., & Drąg, M. (2012). Current and prospective applications of non-proteinogenic amino acids in profiling of proteases substrate specificity. *Biological Chemistry, 393*(9), 843–851.

Kasperkiewicz, P., Poreba, M., Snipas, S. J., Parker, H., Winterbourn, C. C., Salvesen, G. S., & Drag, M. (2014). Design of ultrasensitive probes for human neutrophil elastase through hybrid combinatorial substrate library profiling. *Proceedings of the National Academy of Sciences of the United States of America, 111*(7), 2518–2523.

Lim, K.-H., & Baek, K.-H. (2013). Deubiquitinating enzymes as therapeutic targets in cancer. *Current Pharmaceutical Design, 19*(22), 4039–4052.

McStay, G. P., Salvesen, G. S., & Green, D. R. (2008). Overlapping cleavage motif selectivity of caspases: Implications for analysis of apoptotic pathways. *Cell Death and Differentiation, 15*(2), 322–331.

Mikolajczyk, J., Drag, M., Békés, M., Cao, J. T., Ronai, Z. 'ev, & Salvesen, G. S. (2007). Small Ubiquitin-related Modifier (SUMO)-specific proteases: profiling the specificities and activities of human SENPs. *The Journal of Biological Chemistry, 282*(36), 26217–26224.

Mohamed, M. M., & Sloane, B. F. (2006). Cysteine cathepsins: multifunctional enzymes in cancer. *Nature Reviews. Cancer, 6*(10), 764–775.

Moroy, G., Alix, A. J. P., Sapi, J., Hornebeck, W., & Bourguet, E. (2012). Neutrophil elastase as a target in lung cancer. *Anti-Cancer Agents in Medicinal Chemistry, 12*(6), 565–579.

O'Donoghue, A. J., Jin, Y., Knudsen, G. M., Perera, N. C., Jenne, D. E., Murphy, J. E., Craik, C. S., & Hermiston, T. W. (2013). Global substrate profiling of proteases in human neutrophil extracellular traps reveals consensus motif predominantly contributed by elastase. *PLoS One, 8*(9), e75141.

Perera, N. C., Wiesmüller, K.-H., Larsen, M. T., Schacher, B., Eickholz, P., Borregaard, N., & Jenne, D. E. (2013). NSP4 is stored in azurophil granules and released by activated neutrophils as active endoprotease with restricted specificity. *Journal of Immunology, 191*(5), 2700–2707.

Pham, C. T. N. (2006). Neutrophil serine proteases: Specific regulators of inflammation. *Nature Reviews. Immunology, 6*(7), 541–550.

Pop, C., & Salvesen, G. S. (2009). Human caspases: Activation, specificity, and regulation. *The Journal of Biological Chemistry, 284*(33), 21777–21781.

Poreba, M., & Drag, M. (2010). Current strategies for probing substrate specificity of proteases. *Current Medicinal Chemistry, 17*(33), 3968–3995.

Poreba, M., Gajda, A., Picha, J., Jiracek, J., Marschner, A., Klein, C. D., Salvesen, G. S., & Drag, M. (2012). S1 pocket fingerprints of human and bacterial methionine aminopeptidases determined using fluorogenic libraries of substrates and phosphorus based inhibitors. *Biochimie, 94*(3), 704–710.

Poreba, M., Kasperkiewicz, P., Snipas, S. J., Fasci, D., Salvesen, G. S., & Drag, M. (2014). Unnatural amino acids increase sensitivity and provide for the design of highly selective caspase substrates. *Cell Death and Differentiation, 21*(9), 1482–1492.

Poreba, M., McGowan, S., Skinner-Adams, T. S., Trenholme, K. R., Gardiner, D. L., Whisstock, J. C., To, J., Salvesen, G. S., Dalton, J. P., & Drag, M. (2012). Fingerprinting the substrate specificity of M1 and M17 aminopeptidases of human malaria, Plasmodium falciparum. *PLoS One, 7*(2), e31938.

Poreba, M., Mihelic, M., Krai, P., Rajkovic, J., Krezel, A., Pawelczak, M., Klemba, M., Turk, D., Turk, B., Latajka, R., & Drag, M. (2014). Unnatural amino acids increase activity and specificity of synthetic substrates for human and malarial cathepsin C. *Amino Acids*, *46*(4), 931–943.

Rawlings, N. D., Barrett, A. J., & Bateman, A. (2014). Using the MEROPS database for proteolytic enzymes and their inhibitors and substrates. *Current Protocols in Bioinformatics/Editoral Board, Andreas D. Baxevanis ... [et al.]*, *48*, 1.25.1–1.25.33.

Reyes-Turcu, F. E., Ventii, K. H., & Wilkinson, K. D. (2009). Regulation and cellular roles of ubiquitin-specific deubiquitinating enzymes. *Annual Review of Biochemistry*, *78*, 363–397.

Riedl, S. J., & Salvesen, G. S. (2007). The apoptosome: Signalling platform of cell death. *Nature Reviews. Molecular Cell Biology*, *8*(5), 405–413.

Russell, N. S., & Wilkinson, K. D. (2005). Deubiquitinating enzyme purification, assay inhibitors, and characterization. *Methods in Molecular Biology*, *301*, 207–219.

Salvesen, G. S., & Dixit, V. M. (1997). Caspases: Intracellular signaling by proteolysis. *Cell*, *91*(4), 443–446.

Salvesen, G. S., & Riedl, S. J. (2008). Caspase mechanisms. *Advances in Experimental Medicine and Biology*, *615*, 13–23.

Schilling, O., & Overall, C. M. (2008). Proteome-derived, database-searchable peptide libraries for identifying protease cleavage sites. *Nature Biotechnology*, *26*(6), 685–694.

Thornberry, N. A., Rano, T. A., Peterson, E. P., Rasper, D. M., Timkey, T., Garcia-Calvo, M., Houtzager, V. M., Nordstrom, P. A., Roy, S., Vaillancourt, J. P., Chapman, K. T., & Nicholson, D. W. (1997). A combinatorial approach defines specificities of members of the caspase family and granzyme B. Functional relationships established for key mediators of apoptosis. *The Journal of Biological Chemistry*, *272*(29), 17907–17911.

Turk, B. (2006). Targeting proteases: Successes, failures and future prospects. *Nature Reviews. Drug Discovery*, *5*(9), 785–799.

Wang, F., Krai, P., Deu, E., Bibb, B., Lauritzen, C., Pedersen, J., Bogyo, M., & Klemba, M. (2011). Biochemical characterization of Plasmodium falciparum dipeptidyl aminopeptidase 1. *Molecular and Biochemical Parasitology*, *175*(1), 10–20.

Węglarz-Tomczak, E., Poręba, M., Byzia, A., Berlicki, Ł., Nocek, B., Mulligan, R., Joachimiak, A., Drąg, M., & Mucha, A. (2013). An integrated approach to the ligand binding specificity of Neisseria meningitidis M1 alanine aminopeptidase by fluorogenic substrate profiling, inhibitory studies and molecular modeling. *Biochimie*, *95*(2), 419–428.

Yuan, F., Verhelst, S. H. L., Blum, G., Coussens, L. M., & Bogyo, M. (2006). A selective activity-based probe for the papain family cysteine protease dipeptidyl peptidase I/cathepsin C. *Journal of the American Chemical Society*, *128*(17), 5616–5617.

CHAPTER 5

Shocking Harmony

Magda Dubińska-Magiera and Ewa Błaszczak

Abstract

Harmony is the essence of life. To live, a uni- or multicellular organism must follow a simple rule: maintain a dynamic equilibrium. This means that a cell has to perform all of life's processes, carried out in response to environmental stimuli using various molecules, with proteins being the key players. Under particular environmental conditions, each cell possesses a unique collection of proteins called the proteome. The process whereby a cell controls the synthesis and processing of the proteome is known as proteostasis. A significant feature of proteostasis is harmony – the maintenance of a proteome consisting of correctly folded proteins present at a precise time and place inside the cell. Loss of harmony, or undermining of the cell's dynamic equilibrium, causes problems related to ageing, lack of stress resistance, cancer immunity, and immunity to infections. Due to environmental changes, proteostasis can easily be disturbed. This is why organisms have developed complex mechanisms to carry out one simple requirement: namely, the maintenance of intracellular order. This machinery is referred to as the protein quality control system (PQC). Its basic components are heat-shock proteins (classified as belonging to the group of molecular chaperones), which guard the proteome's integrity.

Keywords

proteome – proteostasis – protein folding – disease states – protein quality control – ubiquitin-proteasome system – heat-shock proteins

•••

> "Harmony"– that's the word that's stuck in my mind. Harmony. It's not about what's lasting or permanent, it's about individual voices coming together for a moment… And that moment lasts the length of a breath.
>
> FRANCIS J. UNDERWOOD (House of Cards, Season 1, Episode 8)

1 Introduction

In 1972, Christian B. Anfinsen, along with Stanford Moore and William H. Stein, received the Nobel Prize in Chemistry for research on the folding (i.e., formation of the spatial structure) of ribonuclease A, an enzyme responsible for the degradation of RNA, one of the nucleic acids. This research concerned the relationship between the activity of a protein and its amino acid sequence and conformation (spatial structure). The outstanding American biochemist proved that a denatured ribonuclease, a ribonuclease with an altered spatial structure, may return to its appropriate conformation and maintain its enzymatic activity. Anfinsen's studies have led to what is referred to as the thermodynamic hypothesis, also known as Anfinsen's dogma, which states that the amino acid sequence of a protein defines its spatial structure. Anfinsen's dogma postulates that in particular conditions needed for proteins to fold, depending, among other requirements, on temperature, solvent composition, and solvent concentration, the native structure of a protein (i.e., the spatial structure that allows the protein to perform its biological functions) is unique and stable. It, however, may be lost due to protein denaturation (Anfinsen, 1973).

Studies focusing on protein folding that have been carried out over many years have confirmed this dogma. However, many exceptions have been discovered that do not meet the criteria of the dogma, e.g., prions and other proteins involved in the formation of pathogenic aggregates (precipitated 'agglomerations' of abnormal proteins) that are related to the development of various neurodegenerative disorders, such as, Alzheimer's disease, Parkinson's disease and Huntington's disease. Furthermore, it has been discovered that, during folding, many polypeptides require the presence of other assisting proteins, called chaperones, to reach their native structure. One of the tasks that chaperones carry out is preventing proteins from aggregating ('sticking together') before reaching their native state (Cox et al., 2020; Hetz & Mollereau, 2014; Higuchi-Sanabria et al., 2018).

How does a protein 'know' how to change a linear sequence of amino acids into a functional structure with a specific spatial orientation? How long does it take for a protein to attain native conformation? These questions have already given biochemists many sleepless nights over the past several decades. The rising interest in protein folding inspired a researcher named Cyrus Levinthal to carry out a thought experiment, now commonly known as Levinthal's paradox. The paradox relates to a considerable discrepancy between the theoretical time needed for a protein to be folded correctly (determined using a mathematical model) and the real-time for a protein to reach its native conformation

(Levinthal, 1969). Assuming that the process takes place through a random sampling of all possible conformations and estimating it using a mathematical model, we can calculate the theoretical amount of time that a protein requires to attain its stable native conformation. The total theoretical time is longer than the supposed age of the Universe! The real-time of folding ranges from milliseconds to seconds. This is because folding is not a linear process; it results from a progressing stabilization of transition forms (Rooman et al., 2002). Protein folding is described with a physical model depicting the concept of minimizing free energy, a theory known as the folding funnel (Dill & Chan, 1997; Sali et al., 1994).

2 Protein: Structure and Functions

Proteins are key components of cells. Alongside nucleic acids, carbohydrates and lipids, they belong to one of the main four classes of macromolecules that constitute the biochemical basis of life. Structurally speaking, proteins are also the most complex and diverse macromolecules. They can be thought of as the 'driving force' of each organism. Their importance is proven by the fact that they directly or indirectly participate in all biological processes (Berg et al., 2002).

The spatial structure of a protein, i.e., protein conformation, depends on the arrangement of amino acids in a polypeptide chain and the chain's manner of folding. A few organization levels can be distinguished in the structure of a protein: primary, secondary, tertiary and quaternary (Figure 5.1). It should be emphasized that despite a strictly determined spatial structure, proteins maintain relative structural flexibility. Thanks to this, they can perform biological functions requiring, for example, local deformations that are necessary for the binding of another molecule (Berg et al., 2002).

3 Proteome: The Protein Pool of a Cell

The set of proteins inside every cell, the proteome, is constantly changing. Maintaining biological equilibrium (homeostasis) requires precise control of the processes that take part in the protein cycle, e.g., the synthesis and degradation of proteins. The state of this equilibrium can be regarded as dynamic. The total pool of proteins at the disposal of a cell may change, for instance, in response to different external signals or due to an illness. These changes cause a new equilibrium to be established (Hinkson & Elias, 2011).

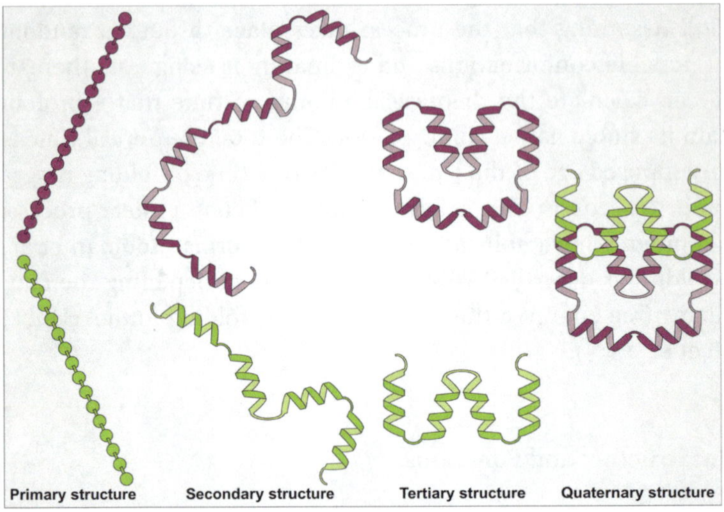

FIGURE 5.1 The types of protein structure. Proteins have four levels of organization: primary, secondary, tertiary, and quaternary structures. The primary structure is a linear sequence of amino acids. The secondary structure refers to three-dimensional forms like alpha-helices or beta-sheets, shaped by weak interactions like hydrogen bonds. The tertiary structure describes the folding of secondary elements into a spatial arrangement through interactions such as hydrogen bonds and disulfide bridges. Quaternary structure arises when multiple folded polypeptide chains combine, though not all proteins have this level

Proteostasis, i.e., protein homeostasis, is the state of equilibrium of the proteome. It encompasses mechanisms responsible for both stabilizing proteins in physiological and stress conditions (mostly involving the participation of chaperone proteins) as well as the mechanisms responsible for the degradation of proteins (involving specialized structures, such as proteasomes and lysosomes, which also engage various chaperone proteins). The protein quality control system (PQC) is responsible for maintaining the integrity of the proteome. The inability to maintain proteostasis causes disorders involving an excessive production, degradation, and accumulation of proteins with an incorrect conformation (Powers & Gierasch, 2021).

4 HSPs: The Basic Elements of the Protein Quality Control System

In 1962, an Italian geneticist, Ferruccio Ritossa, discovered that raising the body temperature of a common fruit fly (*Drosophila melanogaster*) increases the synthesis of numerous proteins. The result of this observation was the very first identification of heat shock proteins (HSPs, in 1974), the name of which refers to the circumstances surrounding their discovery (Ritossa, 1996). Because of their function, they are called assisting proteins or chaperones, and they are 'molecular guardians' for the cell. HSPs are a family of proteins produced by cells in response to exposure to stress conditions. In non-stress conditions, these proteins take part in the regulation of fundamental cellular functions, namely housekeeping. HSPs were first described in the context of the thermal shock that increased their synthesis. However, as it was soon discovered, they may be synthesized through other stress factors, such as free radicals, UV radiation, infections, inflammation and hypoxia (Dubrez et al., 2020; Kregel, 2002; Niforou et al., 2014).

The most popular classification of the family of HSPs was conducted based on their mass, expressed in kilodaltons (kDa). According to this criterion, the family can be divided into six main classes: small heat shock proteins (sHSPs), with a mass between 15 and 30 kDa (Jagla et al., 2018), and HSP40, HSP60, HSP70, HSP90 and HSP100, the names of which denote that the mass of the proteins from these classes equals 40, 60, 70, 90 and 100 kDa, respectively. Cells are usually equipped with many representatives of each class of chaperones (Kampinga et al., 2009).

Chaperones usually work in cycles that involve binding and releasing a substrate, i.e., a protein requiring 'guardianship'. They are regulated by various cofactors, i.e., chemical substances supporting the functioning of proteins. The characteristic feature of most representatives of the HSP family is their activity involving the hydrolysis of high-energy bonds present in adenosine triphosphate (ATP), the substance that 'fuels' most life processes. Thanks to this activity, HSPs obtain the energy necessary to perform their main function, namely, taking part in the folding of other proteins (Martin Vabulas et al., 2010). For example, HSP70 (Figure 5.2) is switched between two states in its cycle: a closed state, with a high affinity for a protein 'client', and an open state, with a low affinity for a protein substrate. The 'client' may be an unfolded or partially folded polypeptide, and the nucleotide exchange factor (NEF) is a cofactor. The NEF takes part in the exchange of adenosine diphosphate (ADP), i.e., ATP after it has been used to obtain energy (Rosenzweig et al., 2019). The HSP70 cycle aims to form the correct structure of proteins belonging to the fast-folding proteins. Polypeptides, which require more time to attain native conformation, use an additional folding system based on HSP60 or HSP90 (Figure 5.3). For them, the HSP70 cycle is a safe transition state, which ensures conditions that prevent potentially toxic aggregation (Saibil, 2013).

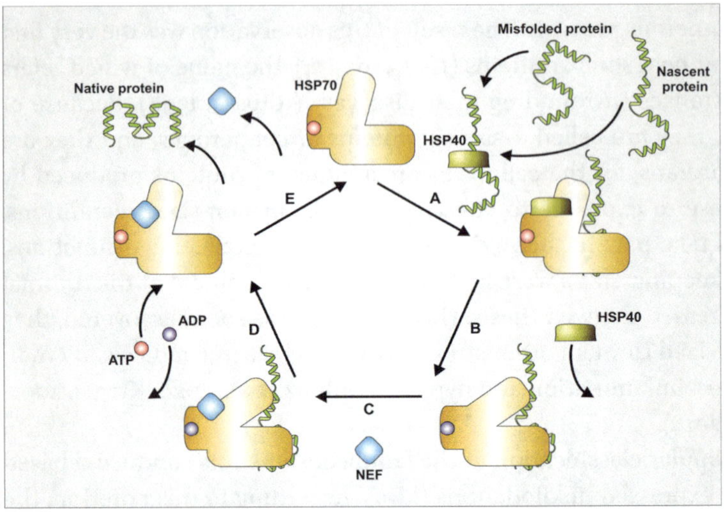

FIGURE 5.2 A simplified diagram of the HSP70 protein's reaction cycle. A. HSP70 binds the substracte, an unfolded or misfolded polypeptide, delivered by an HSP40 chaperone. HSP70 remins in an open state. B. After binding the substrate, HSP40 is released, and ATP is hydrolyzed, providing energy for HSP70 to transition to a closed state. C. The HSP70-ADP-substrate complex binds to a nucleotide exchange factor (NEF), which facilitates the release of ADP and its replacement with ATP, causing HSP70 to return to an open state. D. The folded substrate and NEF are released, allowing the cycle to begin again

Systems based on proteins from the HSP family do not act independently. Instead, they are interconnected and assisted by other proteins that also belong to the family of chaperone proteins. For example, a protein substrate is delivered to the HSP70 system through a protein belonging to the HSP40 class (Rosenzweig et al., 2019). In turn, the repair of damaged polypeptides requires collaboration between the HSP70 and HSP100 or sHSPs systems, depending on the type of damage (Barends et al., 2010; Dubińska-Magiera et al., 2014).

5 The Life Cycle of a Protein

The life cycle of a protein begins with the expression of the gene that encodes the protein, while the degradation of the protein marks the endpoint of its cycle (Figure 5.4). Gene expression is understood as reading the information contained in a gene and transcribing it first into mRNA (messenger RNA) and

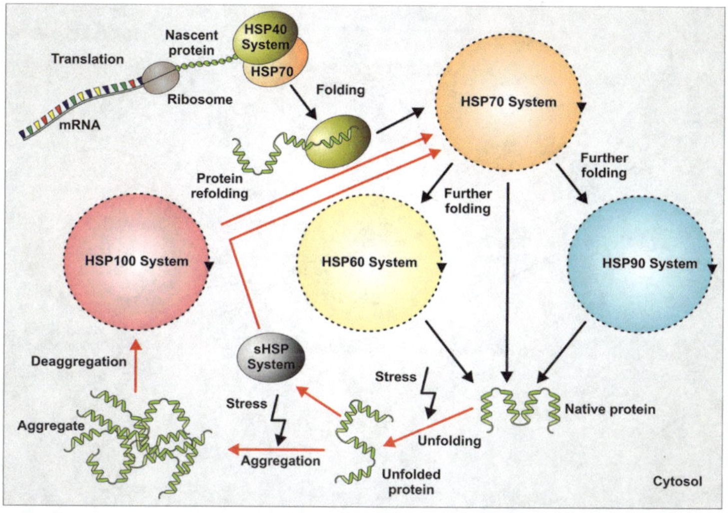

FIGURE 5.3 A simplified diagram of the network of HSP family proteins, the main component of the protein quality control system (PQC). Heat shock proteins (HSPs) are produced by all organisms to maintain proteostasis, with expression levels increased by stress factors. Their main function is to assist in the folding of newly synthesized and denatured polypeptides. HSP70, the primary protein in folding system, collaborates with other chaperone systems like HSP60 and HSP90, mediated by specialized factors not shown in the diagram. Repair systems, such as HSP100, remove protein aggregates and help refold damaged proteins. Small heat shock proteins (sHSPs) prevent irreversible aggregation, directing proteins to repair pathways. Black arrows show protein folding, while red arrows indicate pathways for unfolding, repair, and aggregation triggered by stress factors

then translating it into a protein. Other processes connected with this cycle also encompass folding and transport to cellular compartments and the modification of a protein involving, for example, the addition of a chemical group or the addition of another protein. The appropriate time coordination of these processes is carried out by the elements of the PQC system, primarily the proteins of the HSP family. It is of fundamental importance for the functioning of a cell (Berg et al., 2002).

Translation involves the synthesis of a protein on a ribosome, which decodes information hidden in mRNA and, based on this information, adds specific amino acids, one after another, creating a linear, unbranched polypeptide chain. Folding begins during translation and primarily concerns the formation of secondary structures (Berg et al., 2002; Buchan & Stansfield, 2007). The majority of chaperone proteins that assist other polypeptides in assembling

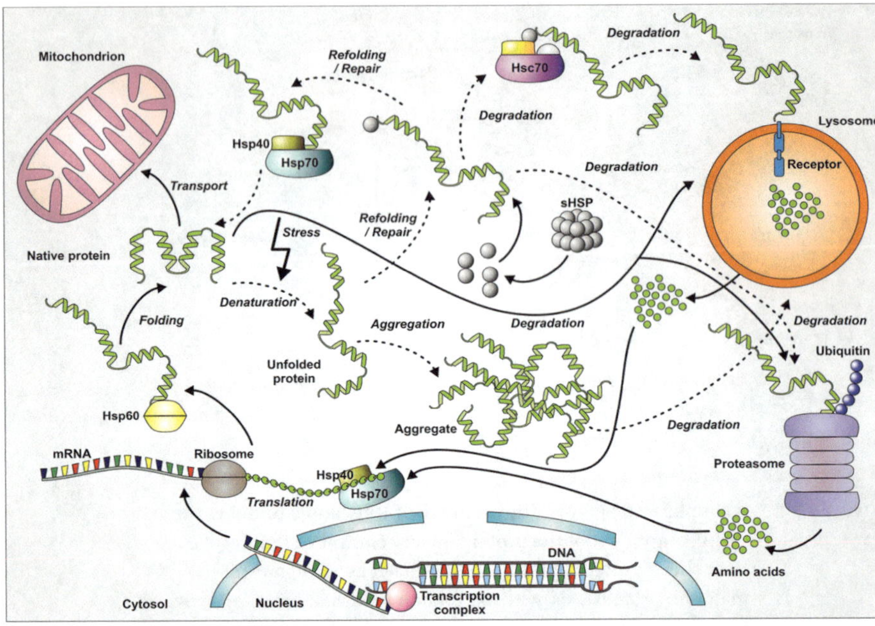

FIGURE 5.4 A simplified diagram of a protein life cycle. During its life cycle, a protein undergoes various processes encompassing the translation of mRNA on a ribosome, folding assisted by chaperones, transport to cellular compartments (such as e.g., a mitochondrion), post-translational modifications and degradation (solid lines). However, under stress conditions (e.g., increased temperature), additional scenarios are possible, e.g., denaturation (the loss of native conformation) and then either repair by re-folding, aggregation, or degradation (dashed lines). The diagram does not include all the aspects of the functioning of chaperone systems and possible degradation pathways. For the sake of clarity, this diagram does not include all PQC systems and the other possible pathways of protein degradation

or disassembling their chains mainly function during the post-translational stages of the protein life cycle. However, specific chaperones begin their work immediately after a sufficiently long fragment of a new polypeptide chain; usually, longer than 60 amino acids, leaves a ribosome. Their involvement mainly concerns promoting interactions between amino acids and between the atoms of a peptide bond. The task of these and many other chaperones also involves binding to, and shielding, hydrophobic amino acids exposed on the surface of the produced protein (Martin Vabulas et al., 2010).

It should be emphasized that the duration of the life cycles of various proteins may be different. Some proteins are physiologically short-lived molecules, and their half-life is measured in minutes. They include regulatory proteins, e.g., p53, which is a transcription factor with the properties of a tumour suppressor. These proteins are rapidly released and rapidly degraded, which is significant for the regulation of key processes, including transcription, translation and cell growth. In turn, other proteins are characterized by a much longer half-life, expressed in days, for instance, the structural proteins of muscles, including myosin and actin. Others, such as collagen and elastin, the proteins responsible for skin elasticity or crystallin, a protein present in the lens of the eye, have their half-lives expressed in years (Glickman & Ciechanover, 2002; Toyama & Hetzer, 2012).

6 Folding

Protein folding is a physical process that involves the folding of a polypeptide chain. As a result of this process, the protein gains a highly organized, stable spatial structure. The correct spatial structure of each protein determines its functions (Berg et al., 2002). It should be emphasized that folding is a complex process prone to errors. Disturbances during the adaptation of the correct spatial structure by a protein usually lead to the synthesis of a molecule with altered properties. These, in extreme cases, might be toxic for the organism and cause, for example, neurodegenerative diseases resulting from the accumulation of incorrect polypeptides (Berg et al., 2002).

Despite the fact that folding is determined mainly by the primary structure of a protein, the process depends on many different factors, such as the presence of water in the environment where folding occurs (cytosol in a cell). Water causes hydrophobic amino acids to display a tendency to 'hide' inside the synthesized protein; in turn, hydrophilic amino acids are located on the protein's surface, where they can interact with surface molecules. As a result, the polypeptide produced is very unstable and requires additional help to arrive at and maintain its native conformation. For many proteins, the requirement for the correct folding of a polypeptide chain is the presence of chaperone proteins, which assist in this process. The folding times of different proteins vary. Usually, small proteins attain their three-dimensional structure very quickly. However, larger molecules require more time and additional assistance from chaperones to reach their native conformation (Hartl & Hayer-Hartl, 2009).

7 Aggregation: Disastrous Consequences for the Cell

The relative structural flexibility of proteins in performing their biological functions also causes their main weakness. Due to various stress factors (e.g., oxidative stress, high temperatures, chemical substances), proteins can easily lose their native conformation (the polypeptide chain gets unfolded). Therefore, proteins also lose their physical, chemical and biological properties. As a result, this leads to the formation of protein aggregates in the cell. An essential and simultaneously dangerous (when speaking of the maintenance of proteostasis) feature of incorrectly folded proteins is their ability to disturb polypeptides with correct conformation. This may lead to the elimination of theoretically functional proteins from the total pool of the proteome by 'trapping' them in toxic aggregates (Ross & Poirier, 2004). Considering the fact that the cellular environment is very 'crowded', it seems justified that a cell decidedly tries to avoid the accumulation of incorrectly folded proteins with a tendency for aggregation. An excess of protein aggregates in the cell is toxic and may lead to cell death and even the death of the entire organism. In humans, over 30 diseases associated with protein aggregation have now been described (Ciechanover & Kwon, 2017). These include neurodegenerative disorders, such as Alzheimer's disease, Parkinson's disease and Huntington's disease, familial amyloid polyneuropathy and amyotrophic lateral sclerosis. Aggregates created by incorrect proteins may be deposited both inside and outside the cell. The latter is often observed in patients with advanced neurodegenerative disorders (Ulrich Hartl, 2017).

The first and most straightforward method of defence against the formation of toxic deposits is the prevention of undesirable aggregation. This is why organisms have developed a complex network of molecular chaperones. Such a network can 'catch' the aggregation-prone transition forms that are created during protein folding and unfolding. Chaperones, including the HSP70 protein can recognize proteins, which due to the unfolding of the polypeptide chain, expose previously 'hidden' hydrophobic amino acids. In the case of failure, organisms have another line of defence that, apart from proteolytic degradation, uses a system based on HSP100 and allows aggregates to be reactivated without the destruction of proteins (Figure 5.3). The HSP100 system temporarily dissolves deactivated proteins through collaboration with the system based on HSP70 (Barends et al., 2010).

Even though cells have developed effective mechanisms that are responsible for both the repair and the removal of damaged proteins, these mechanisms may sometimes fail, e.g., during prolonged stress conditions. In such cases, an option to protect the integrity of the proteome is the sequestration (i.e., capturing and isolating) of inappropriately folded proteins in the spatially separated structures. 'Capturing' incorrectly folded proteins prevents their harmful interference in the biogenesis of newly synthesized polypeptides, which are susceptible to disturbances, thus 'maintaining the order' in the cellular environment (Escusa-Toret et al., 2013; Sontag et al., 2014).

8 Machinery Responsible for the Degradation of Proteins

The machinery responsible for the degradation of proteins encompasses a few different components. The elimination of proteins often results from a cell's response to stress stimuli and leads to reduced protein content in a cell and the release of amino acids. These amino acids may be used again by cells, e.g., during translation (Figure 5.4). Thus, another protein's life cycle, administered by chaperones, may begin. Substrates intended for degradation include, for example, unnecessary proteins that have already fulfilled their biological functions, non-functional proteins, inappropriately folded or unfolded proteins, and protein aggregates. Their degradation occurs with the use of specialized cellular structures, such as proteasomes and lysosomes, which perform the function of 'molecular shredders'. Degradation may occur in several ways, including proteasomal degradation, ER-associated degradation and lysosomal degradation (Ciechanover & Kwon, 2017; Lemberg & Strisovsky, 2021; Soto & Estrada, 2008).

The ubiquitin-proteasome system (UPS) plays a key role in regulating the protein cycle in eukaryotic organisms and thus supports intracellular homeostasis. Its main role involves eliminating incorrectly folded and damaged polypeptides (Figure 5.4) (Toyama & Hetzer, 2012). It is the main system for the degradation of most proteins, particularly proteins with a short half-life (Lilienbaum, 2013). For the protein to be degraded via this system, the polypeptide must be ubiquitylated, which involves attaching ubiquitin molecules to this protein to be recognized by a proteasome (a protein complex with a cylindrical structure comprising many subunits) (Hartl & Hayer-Hartl, 2009). Many structurally related enzymes are involved in the ubiquitylation process, which are specific to various cellular compartments, for instance, the cell nucleus. These enzymes recognize and selectively ubiquitylate particular proteins, the so-called substrates (Glickman & Ciechanover, 2002; Natoli & Chiocca, 2008; Varshavsky, 2012).

Lysosomal degradation encompasses various pathways and involves the degradation of macromolecules in lysosomes, i.e., specialized structures containing numerous hydrolytic enzymes. One such pathway is chaperone-mediated autophagy (CMA) (Figure 5.4) (Kaushik & Cuervo, 2018; Kettern et al., 2010). The CMA is mostly activated as a result of oxidative stress. This pathway contributes to the degradation of soluble proteins, which are equipped with a specific protein motif. This motif is recognized by the heat shock cognate protein 70 (Hsc70), a chaperone that, along with its partners, transports the protein 'client' to the surface of a lysosome. The complex binds to a lysosomal membrane receptor, which allows the complex to be transported inside the lysosome, where the protein substrate is degraded (Bandyopadhyay et al., 2008).

9 PQC System: Collaboration between Components

Eukaryotic cells are divided into various regions that specialize in particular cellular processes. These regions include the cytoplasm, cell nucleus, endoplasmic reticulum, mitochondria and others. For instance, translation, i.e., the process of translating the information contained in a genome into a protein sequence, takes place in the cytoplasm, while other processes related to storing this information and making it available to the cell at the appropriate time occur in the nucleus.

Each cellular compartment also has specific and individual elements of the PQC system (Goldberg, 2003). However, it should be emphasized that despite spatial separation, these systems collaborate closely with one another (Haynes & Ron, 2010; Shibata & Morimoto, 2014). This phenomenon can be easily illustrated using the cell nucleus as an example. The cell nucleus constitutes the isolated but not hermetically closed, 'command centre' of a cell. Due to its physical separation, the nuclear proteome is to a certain extent protected from changes resulting from the conformation disorders of cytoplasmic proteins. The machinery responsible for the protection of the cell's nucleus comprises the components of nuclear PQC (Foresti et al., 2014; Khmelinskii et al., 2014), which form this permanent equipment of this compartment, and also contain some elements of PQC systems that are characteristic of other cellular compartments (Shibata & Morimoto, 2014).

Protein folding disorders are identified by 'detectors' that are specific to the cellular compartment in which they occur. Apart from functioning autonomously, these systems can extend their influence outside the cell, protecting the entire organism from the harmful consequences of proteotoxic stress. The signals that warn against stress may be sent by the cells of other tissues, which can thus participate in protecting other cells located in remote regions of the organism. For instance, stress induced in neurons can protect muscle cells from the consequences of chronic proteotoxicity (Taylor et al., 2014).

The aging of an organism leads to the insufficient functioning of the PQC systems and, consequently, the inappropriately folded or entirely unfolded proteins and the formation of intracellular and extracellular toxic deposits containing 'clustered' proteins and aggregates (Koga et al., 2011). These deposits are manifested in many age-related disorders, such as Alzheimer's disease, Parkinson's disease and cataracts (Ciechanover & Kwon, 2017; Moreau & King, 2012).

10 Conclusions

The Greek origin of the word proteostasis contains the notion of stasis. However, 'stasis' is not used here in the sense of invariability. The term can be considered to mean the dynamic maintenance of equilibrium, i.e., harmony. A cell can maintain its integrity (appearing to remain unchanged) through the numerous perfectly collaborating elements that react to various stimuli from the constantly changing environment. Under the guise of stasis, a fierce fight for the maintenance of equilibrium takes place. To deal with this challenge, cells have to be flexible and systematically verify the quantity and quality of their components, including proteins. This is why keeping the proteome's composition of correctly folded proteins, which are present at a precise time and place inside the cell, is so important for maintaining proteostasis. Broadly defined, stress involves a disturbance to the equilibrium and shifts the internal conditions away from the optimal state. In other words, it causes the loss of harmony. The maintenance of proteostasis is ensured by a complex system of biological pathways where components such as heat shock proteins play the leading role. These proteins supervise the quality and circulation of all polypeptides present in the cell. Thanks to the fact that all of these 'instruments' are perfectly tuned up, the melody of life can be heard in this 'orchestra'.

Acknowledgements

M.D.-M. was funded by the National Science Centre (NCN, Poland), grant no. 2014/13/d/NZ4/02038. E.B. acknowledges the financial support of the National Science Centre (NCN, Poland), grant no. 2016/21/D/NZ1/00285 and the Foundation for Polish Sciences (FNP).

References

Anfinsen, C. B. (1973). Principles that govern the folding of protein chains. *Science, 181*(4096), 223–230.

Bandyopadhyay, U., Kaushik, S., Varticovski, L., & Cuervo, A. M. (2008). The chaperone-mediated autophagy receptor organizes in dynamic protein complexes at the lysosomal membrane. *Molecular and Cellular Biology, 28*(18), 5747–5763.

Barends, T. R. M., Werbeck, N. D., & Reinstein, J. (2010). Disaggregases in 4 dimensions. *Current Opinion in Structural Biology, 20*(1), 46–53.

Berg, J. M., Tymoczko, J. L., & Stryer, L. (2002). *Biochemistry* (5th ed.). W. H. Freeman.

Buchan, J. R., & Stansfield, I. (2007). Halting a cellular production line: Responses to ribosomal pausing during translation. *Biology of the Cell / under the Auspices of the European Cell Biology Organization*, *99*(9), 475–487.

Ciechanover, A., & Kwon, Y. T. (2017). Protein quality control by molecular chaperones in neurodegeneration. *Frontiers in Neuroscience*, *11*, 185.

Cox, D., Raeburn, C., Sui, X., & Hatters, D. M. (2020). Protein aggregation in cell biology: An aggregomics perspective of health and disease. *Seminars in Cell & Developmental Biology*, *99*, 40–54.

Dill, K. A., & Chan, H. S. (1997). From Levinthal to pathways to funnels. *Nature Structural Biology*, *4*(1), 10–19.

Dubińska-Magiera, M., Jabłońska, J., Saczko, J., Kulbacka, J., Jagla, T., & Daczewska, M. (2014). Contribution of small heat shock proteins to muscle development and function. *FEBS Letters*, *588*(4), 517–530.

Dubrez, L., Causse, S., Borges Bonan, N., Dumétier, B., & Garrido, C. (2020). Heat-shock proteins: Chaperoning DNA repair. *Oncogene*, *39*(3), 516–529.

Escusa-Toret, S., Vonk, W. I. M., & Frydman, J. (2013). Spatial sequestration of misfolded proteins by a dynamic chaperone pathway enhances cellular fitness during stress. *Nature Cell Biology*, *15*(10), 1231–1243.

Foresti, O., Rodriguez-Vaello, V., Funaya, C., & Carvalho, P. (2014). Quality control of inner nuclear membrane proteins by the Asi complex. *Science*, *346*(6210), 751–755.

Glickman, M. H., & Ciechanover, A. (2002). The ubiquitin-proteasome proteolytic pathway: Destruction for the sake of construction. *Physiological Reviews*. https://doi.org/10.1152/physrev.00027.2001

Goldberg, A. L. (2003). Protein degradation and protection against misfolded or damaged proteins. *Nature*, *426*(6968), 895–899.

Hartl, F. U., & Hayer-Hartl, M. (2009). Converging concepts of protein folding in vitro and in vivo. *Nature Structural & Molecular Biology*, *16*(6), 574–581.

Haynes, C. M., & Ron, D. (2010). The mitochondrial UPR – protecting organelle protein homeostasis. *Journal of Cell Science*, *123*(Pt 22), 3849–3855.

Hetz, C., & Mollereau, B. (2014). Disturbance of endoplasmic reticulum proteostasis in neurodegenerative diseases. *Nature Reviews. Neuroscience*, *15*(4), 233–249.

Higuchi-Sanabria, R., Frankino, P. A., Paul, J. W., 3rd, Tronnes, S. U., & Dillin, A. (2018). A futile battle? Protein quality control and the stress of aging. *Developmental Cell*, *44*(2), 139–163.

Hinkson, I. V., & Elias, J. E. (2011). The dynamic state of protein turnover: It's about time. *Trends in Cell Biology*, *21*(5), 293–303.

Jagla, T., Dubińska-Magiera, M., Poovathumkadavil, P., Daczewska, M., & Jagla, K. (2018). Developmental expression and functions of the small heat shock proteins in Drosophila. *International Journal of Molecular Sciences*, *19*(11), 3441.

Kampinga, H. H., Hageman, J., Vos, M. J., Kubota, H., Tanguay, R. M., Bruford, E. A., Cheetham, M. E., Chen, B., & Hightower, L. E. (2009). Guidelines for the nomenclature of the human heat shock proteins. *Cell Stress & Chaperones*, *14*(1), 105–111.

Kaushik, S., & Cuervo, A. M. (2018). The coming of age of chaperone-mediated autophagy. *Nature Reviews. Molecular Cell Biology*, *19*(6), 365–381.

Kettern, N., Dreiseidler, M., Tawo, R., & Höhfeld, J. (2010). Chaperone-assisted degradation: Multiple paths to destruction. *Biological Chemistry*, *391*(5), 481–489.

Khmelinskii, A., Blaszczak, E., Pantazopoulou, M., Fischer, B., Omnus, D. J., Le Dez, G., Brossard, A., Gunnarsson, A., Barry, J. D., Meurer, M., Kirrmaier, D., Boone, C., Huber, W., Rabut, G., Ljungdahl, P. O., & Knop, M. (2014). Protein quality control at the inner nuclear membrane. *Nature*, *516*(7531), 410–413.

Koga, H., Kaushik, S., & Cuervo, A. M. (2011). Protein homeostasis and aging: The importance of exquisite quality control. *Ageing Research Reviews*, *10*(2), 205–215.

Kregel, K. C. (2002). Invited review: Heat shock proteins: modifying factors in physiological stress responses and acquired thermotolerance. *Journal of Applied Physiology*, *92*(5), 2177–2186.

Lemberg, M. K., & Strisovsky, K. (2021). Maintenance of organellar protein homeostasis by ER-associated degradation and related mechanisms. *Molecular Cell*, *81*(12), 2507–2519.

Levinthal, C. (1969). *How to fold graciously*. https://www.semanticscholar.org/paper/1ef89dfb1e3404f4ace99399ce582b2bc982d0bf

Lilienbaum, A. (2013). Relationship between the proteasomal system and autophagy. *International Journal of Biochemistry and Molecular Biology*, *4*(1), 1.

Martin Vabulas, R., Raychaudhuri, S., Hayer-Hartl, M., & Ulrich Hartl, F. (2010). Protein folding in the cytoplasm and the heat shock response. *Cold Spring Harbor Perspectives in Biology*, *2*(12), a004390.

Moreau, K. L., & King, J. A. (2012). Protein misfolding and aggregation in cataract disease and prospects for prevention. *Trends in Molecular Medicine*, *18*(5), 273–282.

Natoli, G., & Chiocca, S. (2008). Nuclear ubiquitin ligases, NF-κB degradation, and the control of inflammation. *Science Signaling*, *1*(1), e1–pe1.

Niforou, K., Cheimonidou, C., & Trougakos, I. P. (2014). Molecular chaperones and proteostasis regulation during redox imbalance. *Redox Biology*, *2*, 323–332.

Powers, E. T., & Gierasch, L. M. (2021). The proteome folding problem and cellular proteostasis. *Journal of Molecular Biology*, *433*(20), 167197.

Ritossa, F. (1996). Discovery of the heat shock response. *Cell Stress & Chaperones*, *1*(2), 97–98.

Rooman, M., Dehouck, Y., Kwasigroch, J. M., Biot, C., & Gilis, D. (2002). What is paradoxical about levinthal paradox? *Journal of Biomolecular Structure & Dynamics*, *20*(3), 327–329.

Rosenzweig, R., Nillegoda, N. B., Mayer, M. P., & Bukau, B. (2019). The Hsp70 chaperone network. *Nature Reviews. Molecular Cell Biology, 20*(11), 665–680.

Ross, C. A., & Poirier, M. A. (2004). Protein aggregation and neurodegenerative disease. *Nature Medicine, 10*(7), S10–S17.

Saibil, H. (2013). Chaperone machines for protein folding, unfolding and disaggregation. *Nature Reviews. Molecular Cell Biology, 14*(10), 630–642.

Sali, A., Shakhnovich, E., & Karplus, M. (1994). How does a protein fold? *Nature, 369*(6477), 248–251.

Shibata, Y., & Morimoto, R. I. (2014). How the nucleus copes with proteotoxic stress. *Current Biology: CB, 24*(10), R463–R474.

Sontag, E. M., Vonk, W. I. M., & Frydman, J. (2014). Sorting out the trash: the spatial nature of eukaryotic protein quality control. *Current Opinion in Cell Biology, 26*, 139–146.

Soto, C., & Estrada, L. D. (2008). Protein misfolding and neurodegeneration. *Archives of Neurology, 65*(2), 184–189.

Taylor, R. C., Berendzen, K. M., & Dillin, A. (2014). Systemic stress signalling: Understanding the cell non-autonomous control of proteostasis. *Nature Reviews. Molecular Cell Biology, 15*(3), 211–217.

Toyama, B. H., & Hetzer, M. W. (2012). Protein homeostasis: Live long, won't prosper. *Nature Reviews. Molecular Cell Biology, 14*(1), 55–61.

Ulrich Hartl, F. (2017). Protein misfolding diseases. *Annual Review of Biochemistry, 86*, 21–26.

Varshavsky, A. (2012). The ubiquitin system, an immense realm. *Annual Review of Biochemistry, 81*, 167–176.

CHAPTER 6

Can Consonance Be Explained with Mathematical Proportions?

The Simplicity and Complexity of Harmonic Relations in Music

Adrian Foltyn

Abstract

The article focuses on the ideas of simplicity and complexity used to explain musical consonance and dissonance. Selected historical philosophical and mathematical concepts are compared with experimental research on the perception of music intervals. Multiple explanations of these phenomena in the scientific literature are not fully consistent with music practice. The author outlines a novel model based on synchronization of the cognitive system's response with the frequency of incoming sounds. The model utilises concepts of harmonic and prime numbers to represent the cognitive simplicity and stability of musical interval proportions.

Keywords

neurocognitive science – ratio – proportion – consonance – dissonance – harmony – harmonic entropy

1 Introduction

According to available records, the relations between music pitches have been the subject of keen interest to philosophers and mathematicians for the past two and a half thousand years (Tenney, 1988). The Pythagoreans sought references to the harmony of spheres in those relations, manifesting in the perfect, inaudible music of the celestial bodies. The secret of perfect proportions has since then inspired researchers looking for a simple and consistent explanation of the world around us, a kind of *theory of everything*.

DOI:10.1163/9789004714540_006

Numerous speculative theories have been proposed, combining musical harmony with mathematical proportions. The common theme of these theories is the simplicity of ratios corresponding to the perception of musical intervals. In this article, selected definitions of proportion simplicity will be confronted with music practice. The aim is to answer the question of whether the simplicity or complexity of mathematical ratios between pitches could be a determinant of their consonance degree.

2 Definitions

In the following discussion, pitch is referred to as the attribute of auditory sensation by which sounds are ordered on the scale used for melody in music (ASA, 2023). Hence, the relation of two pitches, represented by the frequency ratio of two or more tones played on musical instruments, denotes a mathematical proportion. The pitches can interact in a vertical or horizontal relation, corresponding to harmonic or melodic intervals, respectively.

Importantly, the proportions are derived from the so-called just tuning of musical intervals, an integer-ratio system described by Claudius Ptolemy in the 2nd century BC. Over the recent three centuries, the adjusted 'equally tempered' tuning has become widely used, however small deviations from just intervals do not change the perception of consonance, as already postulated by Aristoxenus in the 3rd century BC. Further, I will ignore phenomena related to the irregular (non-harmonic) spectrum of frequencies, as well as effects observed for very low or very high sounds, i.e., natural roughness or the decreasing accuracy of pitch perception above 2500 Hz (Langner, 2007), respectively. To exemplify the argument, I will only consider the proportions within one octave, i.e., between the ratios of 1:1 and 2:1.

Consonance will be herewith defined as a sound of two or more pitches that listeners consider to be in concordance with each other, evoking the sensations of fusion and stability. The opposite of consonance is dissonance, denoting an unconcordant, rough and unstable sound, the accumulation of which is regarded as less pleasant to listen to. Studies on human perception of musical intervals (e.g., Malmberg, 1918; Bowling et al., 2018) agree in this regard with the theory of the counterpoint, or the art of combining musical voices (Tinctoris, 1477), which assumes the following classification of musical intervals:

- consonances are considered to include the octave (12 semitones), perfect fifth (7), perfect fourth (5), major and minor third (4 and 3), major and minor sixth (9 and 8);

– dissonances are considered to include major and minor seconds (2 and 1), major and minor sevenths (11 and 10) and the tritone (6).

3 Historical Context

The Pythagoreans defined simplicity of proportions as the relation between the smallest natural numbers that formed the so-called *tetraktys* – the Arch-Four. The resulting proportions determined perfect consonances: 2:1 (octave), 3:2 (perfect fifth) and 4:3 (perfect fourth). All other proportions were considered dissonances.

Medieval and Renaissance theorists, starting with an anonymous thirteenth-century theorist and followed by Johannes de Garlandia, Walter Odington and Bartolomeo Ramis de Pareia, gradually acknowledged the role of thirds and even sixths as consonances. Thus, not only the proportions formed by the first four but also by further small natural numbers, came to be treated as consonances. The *tetraktys* ceased to be the determinant of proportion simplicity, replaced by the so-called *senario* (Zarlino, 1558), i.e. a series of natural numbers up to 6. Consequently, with the development of musical harmony and growing composer preference towards the major triad (4:5:6), consonances were extended to the major sixth (5:3) as well as the major and minor third (5:4 and 6:5).

Over time, the progress of the major-minor system meant that the minor sixth was added to the consonances, even though it did not fit into the *senario* with its 8:5 ratio. The inclusion of this interval, however, was natural as the minor sixth is an inversion (complement of the octave) of another interval already regarded as a consonance, i.e. the major third (Table 6.1).

TABLE 6.1 Musical intervals and their corresponding simple ratios in just intonation

Interval	Number of semitones	Ratio	Interval	Number of semitones	Ratio
Minor second	1	16:15	Perfect fifth	7	3:2
Major second	2	9:8	Minor sixth	8	8:5
Minor third	3	6:5	Major sixth	9	5:3
Major third	4	5:4	Minor seventh	10	9:5
Perfect fifth	5	4:3	Major seventh	11	15:8
Tritone	6	7:5	Perfect octave	12	2:1

FIGURE 6.1 Geometric representation of simple ratios by Descartes

Descartes attempted to represent the known musical intervals using geometrical proportions to explain proportion simplicity and consonance (Sepper, 1996) (Figure 6.1).

The points C, D, E and F bisect (divide in half) sections AB, BC, CD and CE, respectively. According to Descartes, only the proportions resulting from points A to E create consonances, while proportions containing point F form dissonances. This theory, however, suffers from inconsistencies, as it does not explain the major sixth's 5:3 ratio, being one of the key consonances, while point F can also be included in the consonant proportion of AD:AF (4:3).

Benedetti (1585) suggested that the acoustic properties of harmonic sounds and the total period duration of musical intervals formed by them are related to simple natural number ratios and thereby constitute an explanation of consonance. According to Benedetti, the total period of an interval's acoustic representation was proportional to the product of the numerator and the denominator of the corresponding ratio, expressed as an irreducible fraction. However, the minor sixth (8:5) remained a problem, because there exists a 'simpler' proportion – composed of smaller natural numbers – evoking a dissonance sensation. The problematic ratio is 7:5, equal approximately to the tritone, one of the strongest dissonances, dubbed 'the devil in music" by counterpoint theorists. In conclusion, proportions formed by the smallest natural numbers cannot be directly equated with consonance.

A further innovative concept was put forth by the famous mathematician Leonard Euler and termed *gradus suavitatis* (Latin: *degree of softness*; Bailhache, 1997). Euler proposed a mathematical formula for the simplicity of proportions, based on the factorisation of the product of numbers representing a proportion. The formula was:

$$G = 1 + \Sigma_{i=1} k_i \left(p_i - 1\right) = 1 + \Sigma_{i=1} k_i \left(p_i - 1\right) \tag{1}$$

where p stood for the factor, and k for its power.

For example, the factorisation of the minor sixth's ratio (8:5) is $2^3 * 5^1$, which, based on (1), results in the gradus suavitatis: $G = 1 + 3\,(2 - 1) + 1\,(5 - 1) = 8$.

Notably, it was the first known theory that tackled the abovementioned problem of the tritone and the minor sixth (cf. Table 6.2). Unfortunately, the concept displays other shortcomings, such as a similar consonance degree of the minor third (6:5), the minor sixth (8:5) and the minor second (9:8). All of them are placed in the eighth gradus, which is at odds with the dominant perception of the first two intervals as consonances and the third one as a dissonance. Therefore, the simplicity of ratios proposed by Euler cannot be considered as an exhaustive explanation of consonance.

In the following centuries, studies in acoustics and signal processing by the human auditory system led to a better understanding of consonance and dissonance rooted in physical phenomena (von Helmholtz, 1863). Bohlen (1978) attempted to combine the simplicity of ratios with the phenomenon of combination tones, but the resulting consonance measures showed the same problem that had already manifested itself in Benedetti's work.

An intriguing and consistent mathematical concept of consonance, based on the Farey series, was proposed by Erlich (1997). The Farey series of the F-th order includes all fractions with the numerator and denominator not larger than F. Using terms from the information theory, Erlich postulates that the dissonance sensation increases with the so-called harmonic entropy, defined as the cumulative probability of perceiving a musical interval as a mistuned version of another interval (ratio). The Farey series proves useful in this case, as one of its properties is a larger distance between its neighbouring terms for small-number ratios. This demonstrates lower 'uncertainty' of the cognitive system when dealing with simple, consonant intervals (Figure 6.2).

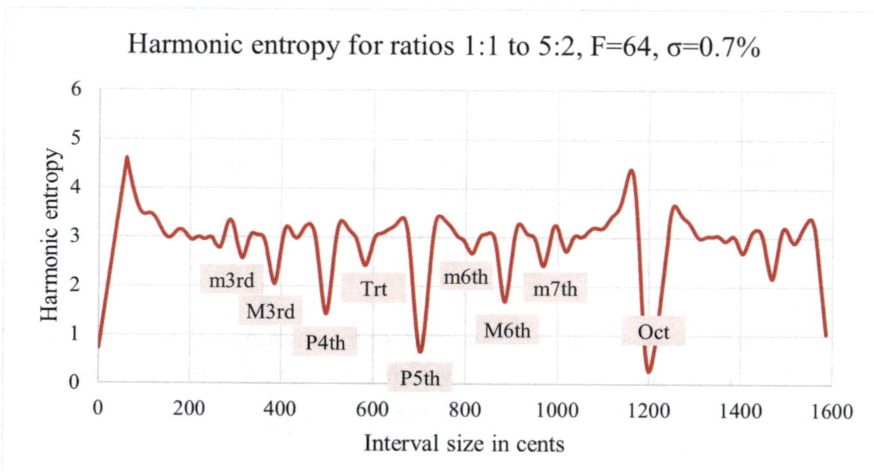

FIGURE 6.2 Measure of harmonic entropy of order F = 64 for intervals between the unison and the octave + major third. Common intervals from 'deeps' in the graph and are marked accordingly. Notation of intervals: m – minor, M – major, P – perfect, Trt – tritone, Oct – octave. Cents (c) denote the proportion reflecting an interval as $2^{c/1200}:1$

Closer analysis reveals that the harmonic entropy of the tritone (7:5) is quite low compared to the remaining intervals, including those commonly perceived as consonances. Yet, the hypothesis, that there is a certain cognitive complexity associated with the sensation of dissonance, is likely reasonable and reflects not only in the perceived degree of simplicity (ease) but also the stability of music intervals.

4 Hypothesis

Recent evidence suggests that the human cognitive system responds to the frequency of the incoming signal in a synchronised manner. This phenomenon has been observed at different stages of the so-called auditory pathway, i.e. the route of nerve impulses representing acoustic stimuli in the brain (Langner, 2007). Assuming that the cognitive system synchronises with the incoming signal, each period of the sound wave generates a (generalized) impulse, or discharge.[1] Importantly, there appears to be a separate cognitive axis responding to pitch (as an aggregate of incoming frequencies of a harmonic tone), on top of known responses to individual frequencies. Moreover, there exist 'specialized' units of the system, responding only to a particular pitch and its corresponding frequency. The mathematics of periodic signals provides an important observation here: a signal with the period (inverse of frequency) of p also 'contains' the periods of $2p$, $3p$, etc. Therefore, the signal with a particular period elicits responses of cognitive units (neuron groups) tuned to that period as well as to its integer multiples (Figure 6.3).

With regard to the conditions presented above, I have proposed the following mechanism:

- When two or more harmonic tones are heard simultaneously, the impulses (i.e., the model responses) form a specific pattern that can be analysed by the cognitive system.
- This pattern is repeated (or repeated approximately) in periods called *pattern windows* and the cognitive system is able to measure the duration of those periods by analysing the time between repeating patterns.
- Following Euler's and Erlich's intuitions, the perceived consonance degree is a result of pattern simplicity and stability. Both metrics represent the ease of processing a given pattern, and in turn, the corresponding ratio and musical interval, by the human cognitive system.

Pattern simplicity is naturally connected with the volume of model responses in a pattern. Its formal representation is the reciprocal of the average number of model responses in a pattern window. It can be conveniently

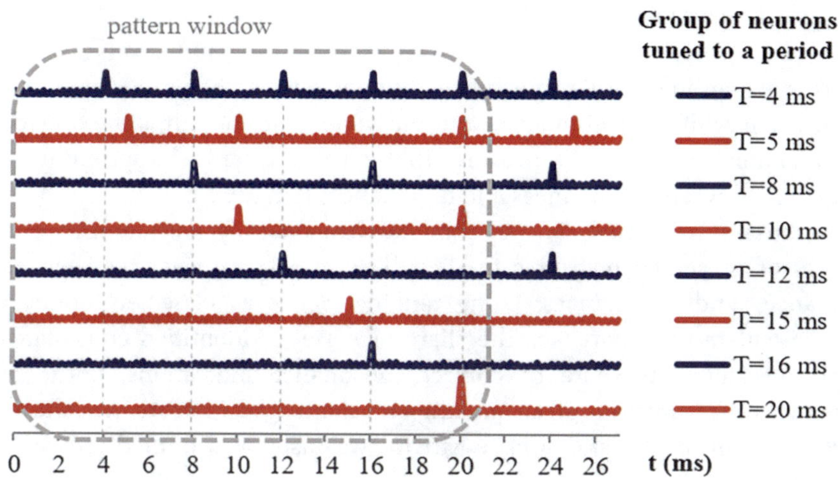

FIGURE 6.3 Pattern window of model responses to two signals with periods of 4 ms and 5 ms (250 Hz and 200 Hz), forming the major third interval. The window length is 20 ms, and it approximately repeats every 20 ms. Note the responses of cognitive units (neuron groups) 'tuned' to multiples of the signal periods

expressed using so-called harmonic numbers, i.e., sums of reciprocals of natural numbers.

$$PSi_{p:q} = \frac{2}{H_p + H_q} \tag{2}$$

where $p{:}q$ is the initial ratio and H_k is the sum of reciprocals of natural numbers up to k.

Pattern stability describes the similarity degree of successive pattern repetitions. The most stable components of a pattern are formed by responses of units tuned to periods equal to integer divisions of the pattern window length. In the case of Figure 6.3 and the pattern window length of 20 ms, these are units tuned to periods of 4 ms, 5 ms and 10 ms. Formally:

$$PSt_{p:q} = \frac{p\sum_{m=1}^{p}\frac{1}{LCM(p,m)} + q\sum_{n=1}^{q}\frac{1}{LCM(q,n)}}{p+q} \tag{3}$$

where LCM is the least common multiple.

Consequently, for a given mathematical proportion $p{:}q$, representing the respective musical interval, the product of pattern simplicity and stability forms the measure of consonance degree.

5 Discussion

It is worth noting that for intervals within one octave, the proposed metric is consistent with musical practice and theoretical intuition. It is the only concept among the described measures that makes a clear and appropriate distinction between consonances and dissonances (Table 6.2).

A further meta-analysis was conducted to quantify and confirm differences of presented measures in their discriminatory power between consonances and dissonances. Using ranking values as classifier inputs, we can calculate sensitivity, i.e., the share of correctly identified consonances, and precision, i.e. share of correct consonance indications, both computed with regard to a given ranking threshold. The F1 score, i.e., the harmonic mean of precision and sensitivity, was used as a unified measure of discriminatory power.

Results confirm that the proposed measure was the only one out of the examined metrics that – in alignment with experimental results – achieved an F1 score equal to 1 (Figure 6.4).

TABLE 6.2 Comparison of selected measures of consonance in terms of distinction between consonances and dissonances

Interval	Musical/ theoretical classification	Ranking by Benedetti's measure	Ranking by Euler's measure	Ranking by Erlich's measure	Ranking by the author's measure	Ranking from experimental results
Minor third	Consonance	6	6	8	6	6
Major third	Consonance	5	4	5	4	3
Perfect fourth	Consonance	3	3	3	3	5
Perfect fifth	Consonance	2	2	2	2	2
Minor sixth	Consonance	8	6	9	7	6
Major sixth	Consonance	4	4	4	5	4
Perfect octave	Consonance	1	1	1	1	1
Minor second	Dissonance	12	11	12	12	12
Major second	Dissonance	10	6	10	8	9
Tritone	Dissonance	7	11	6	9	8
Minor seventh	Dissonance	9	9	7	10	9
Major seventh	Dissonance	11	10	11	11	11

Note: Red markings denote misalignment with musical practice. Experimental results were obtained as the average rank from Malmberg (1918) and Bowling et al. (2018).

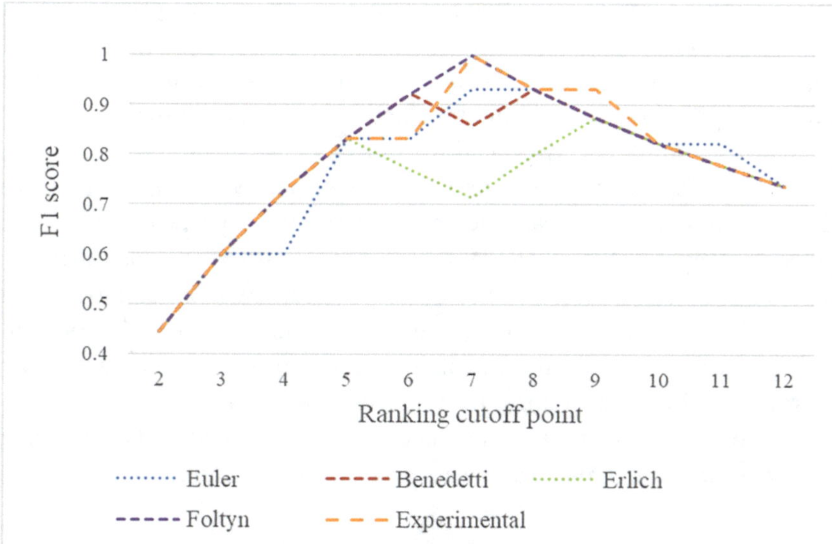

FIGURE 6.4 The F1 score computed for four described measures and experimental results, based on ranking thresholds discriminating dissonances and consonances (see Table 6.2)

This value denotes perfect discrimination between consonances and dissonances within one octave. We can therefore conclude that the suggested mechanism could potentially contribute to the formation of consonance and dissonance sensation.

As a further noteworthy fact, the measures of consonance/dissonance selected for this study represent only a small fraction of to-date mathematical, musical, psychoacoustic and neurophysiological concepts explaining consonance and dissonance (Foltyn, 2012). Curiously, however, multiple theories result in the same rank order of intervals as in the presented concepts. For example, interval orderings based on Kuile (1914), Bohlen (1978), Ebeling (2007), Gill & Purves (2009) and Trulla et al. (2018) coincide exactly with Benedetti's ranking, thereby displaying the same advantages and shortcomings.

Reverting to the original idea of cognitive simplicity, there is only partial correspondence between small natural numbers constituting proportions of musical intervals and their perception as consonances. Further factors can impact perceived interval stability, for instance primeness of numbers forming a ratio. In line with (3), higher prime numbers result in larger LCMs, lower pattern stability, and so, lower consonance degree. This might explain why the 7:6 ratio (minor third reduced by approximately a quartertone, producing the *blue note* popularized by jazz music) is perceived as more consonant than the 7:5

tritone, a smaller-number ratio, but consisting of two highest primes among one-digit natural numbers.

It appears then that the simplicity of a musical interval's proportion is only a prerequisite and not a sufficient condition for the consonance sensation to occur. Computational complexity associated with processing signals and their corresponding ratios depends on the pattern of neural responses it elicits, turning our attention to other mathematical concepts like harmonic or prime numbers. This idea sheds new light on the past and future quests of mathematicians and music theorists for an objective measure of consonance. Clearly, without understanding the mechanisms of processing pitch information by the human cognitive system it is difficult to speculate about generalised mathematical principles underlying these processes. However, tempting the vision of a single simple formula explaining the entire perception of consonance and dissonance might be, it is becoming increasingly evident that reality is at least partially more complex.

Note

1 This is not true for single auditory nerve fibres or neurons in further parts of the auditory pathway, however, neurons respond in groups and synchronise partly independently of each other. Hence, the final information available to the auditory cortex probably includes regular neural firings synchronised with each signal period.

References

ASA: Acoustical Society of America. (2023). *Pitch*. Retrieved October 15, 2023, from https://asastandards.org/terms/pitch

Bailhache, P. (1997). *Music translated into mathematics: Leonhard Euler*. Retrieved June 10, 2023, from http://www.tonalsoft.com/sonic-arts/monzo/euler/euler-en.htm

Benedetti, G. B. (1585). *Diversarum Speculationum mathematicarum et physicarum liber*. Taurini: apud haeredem N. Bévilaquae.

Bohlen, H. (1978). 13 Tonstufen in der Duodezime. *Acustica, 39*, 76–86.

Bowling, D. L., Purves, D., & Gill, K. Z. (2018). Vocal similarity predicts the relative attraction of musical chords. *Proceedings of the National Academy of Sciences of the United States of America, 115*, 216–221.

Ebeling, M. (2007). *Verschmelzung und neuronale Autokorrelation als Grundlage einer Konsonanztheorie*. Frankfurt: Peter Lang.

Erlich, P. (2004). On harmonic entropy. In J. Monzo (Ed.), *Encyclopedia of tuning*. Retrieved June 10, 2023, from http://www.tonalsoft.com/enc/e/erlich/harmonic-entropy_with-commentary.aspx

Foltyn, A. (2012). *Neuroscientific measure of consonance.* Materials from the 2012 ICMPC in Thessaloniki.

von Helmholtz, H. (1863). *Die Lehre von den Tonempfindungen als physiologische Grundlage fuer die Theorie der Musik.* Braunschweig: F. Vieweg & Sohn.

Kuile, T. E. ter. (1914). Konsonanz und einfaches Zahlenverhältnis. *European Journal of Physiology, 159*(1–3), 35–50.

Langner, G. (2007). Die zeitliche Verarbeitung periodischer Signale im Hörsystem: Neuronale Repräsentation von Tonhöhe, Klang und Harmonizität. *Zeitschrift Audiologie, 46*(1), 8–21.

Malmberg, C. F. (1918). The perception of consonance and dissonance. *Psychological Monographs, 25*(2), 93–133.

Sepper, D. L. (1996). *Descartes's imagination: Proportion, images, and the activity of thinking.* Berkeley, CA: University of California Press.

Tenney, J. (1988). *A history of 'consonance' and 'dissonance'.* New York, NY: Excelsior Publishing Company.

Tinctoris, J. (1477). *The art of counterpoint* (A. Seay, Trans.). American Institute of Musicology.

Trulla, L. L., Di Stefano, N., & Giuliani, A. (2018). Computational approach to musical consonance and dissonance. *Frontiers in psychology, 9*, 381.

Zarlino, G. (1558) Istituzioni armoniche (Book 3). In O. Strunk (Ed.), *Source readings in music history.* New York, NY: Norton.

CHAPTER 7

Reinforcement Learning Tasks

From Complexity to Simplicity

Dorota Frydecka and Jarosław Drapała

Abstract

In the field of computational psychiatry, reinforcement learning tasks aim to measure a subject's sensitivity to rewards and punishments. The analysis of behavioral data acquired from human subjects undergoing the task may give an insight into the decision-making process. Despite the complexity of the process, very simple computational models can deliver a concise summary of the subject's performance, this being reduced to a few numbers representing such measures as sensitivity to rewards and punishments or the inclination to explore and exploit. On the other hand, researchers employ complex models, such as biologically plausible neural networks, to furnish a mechanistic account of the experimental data by mimicking the neural circuits involved in decision-making. Hence, the starting point is a piece of complex brain machinery performing a series of decisions, with the time-series data from this simplified to a few parameters pertaining to the computational models used to understand the process "under the hood" – and, finally, the artificial neural network is used to simulate the complex process of decision-making that takes place in the human brain.

Keywords

computational psychiatry – reinforcement learning tasks – neural network model – computational model

1 Introduction

The world around us is full of complex processes that we used to classify as physical, biological, psychological, social, and so on. The key point is that if we take a closer, analytical look, we may find simple rules driving these processes at a lower level of organization. For instance, the complex behaviour of physical objects can be driven by a few simple principles of Newtonian dynamics. The complex universe of chemical reactions may be explained in great part by a very basic description of the interaction between an electron and a photon. The fundamental principles of dynamics in chaotic systems lead to complex behavior in these systems, such as fluid movement, population growth and decay, or weather changes. Another example is the evolution process, which can be reduced to a few simple principles, resulting in the so-called directed random walk. Evolution can generate extremely intricate and organized creations: biological creatures or artificial "organisms" in computer memory (such as artificial animals in the Framsticks project (*3D Simulation and Evolution*, n.d.)). The basic rules of the "Prisoner's Dilemma" game are simple, yet they can explain complex social behavior, such as the spontaneous emergence of altruistic behavior in a society of egoists (Ridley, 1994).

In computational neuroscience, scientists aim to understand the neurobiological foundation of human psychological functioning by constructing and analyzing models from two general categories: *mathematical computational models* and *biologically inspired neural networks*. These approaches bridge the gap between the biophysical level of analyzing the features of individual neurons and their networks (morphological structure of individual neurons, permeability of ion channels, intracellular signalling cascades, synaptic plasticity) and the level of mental phenomena (perception, learning, decision-making, executive or linguistic functions). In this article, we will look at how simplified mathematical models may help us comprehend processes like learning and decision-making.

The term *understand* has two different meanings when studying a complex system. The first might be thought of as a bottom-up perspective: given a set of underlying rules governing a certain degree of system organization, deduce the higher-level implications of these rules. In contrast, a top-down perspective would be: *guess* the lower-lever rules that operate behind the curtain based on higher-level observations of system behavior. We emphasize the word *guess* because that is primarily what the computational neuroscience community does. Researchers take behavioral data from people doing a cognitive task (for example, learning from rewards and punishment), and combine it with expert knowledge of how brain processes deal with cognitive tasks to try to establish the principles that the nervous system follows to complete the tasks. The rules are represented by mathematical models, some of which are simple while others are quite intricate. The simpler the model, the easier it is to understand. The more sophisticated the model, the more chance that it fits to the data.

This is the dilemma that the scientific community has been struggling with for centuries.

The goal of our research is to investigate the interplay between the conscious/implicit level of brain processing and the explicit/unconscious level. As a test bed, we employ a computer game that allows us to collect behavioral data of people's making decisions under uncertainty. Mathematical models account for the explicit level. The game is described in detail in the section that follows. Then, we outline two modeling approaches: one using a complex neural network model and the other using a simple computational model.

2 Probabilistic Learning Task

In brief, a subject is asked to choose between two options: one is better, one is worse. When the game begins, the subject does not know which one is better. Every choice may be rewarded or punished, no matter which option is chosen, but it is more likely to get a reward for picking a better option. The task is to learn which one is better during the trial-and-error process. If a subject's ability to learn from experience and his decision-making skills are not impaired, he will tend to choose the option that gives us rewards more often. Figure 7.1 depicts a single trial pass. The game is referred to as a probabilistic learning task due to randomized payouts (Doll et al., 2009).

There is also a variation of the game, when a subject is given a misleading hint regarding which option is better (Doll et al., 2009). This is an example of information being processed on a direct/conscious level. The rate at which a person unlearns a misleading cue provides some insight into the relationship between the explicit and implicit levels of information processing in the brain.

Now we will go through the game details, where options are symbols displayed on a computer screen. The symbols are Japanese Hiragana characters, which most players are unlikely to have encountered before and – as a result – are equally difficult for anyone to remember and no one should have any previous associations with these characters. The presentation of a pair of symbols is preceded by the display of a green circle for a random time between 1 and 3 seconds. During the presentation of the symbols, the player has 3 seconds to make a choice. She/he then gets a feedback message of 'Good or 'Bad'. Approximately two hundred trials are passed per test.

A probability of receiving a reward p, with a probability of receiving a penalty equal to $1 - p$ is associated with each symbol. The p probabilities are not known to the player. One learning element is estimating their value from the

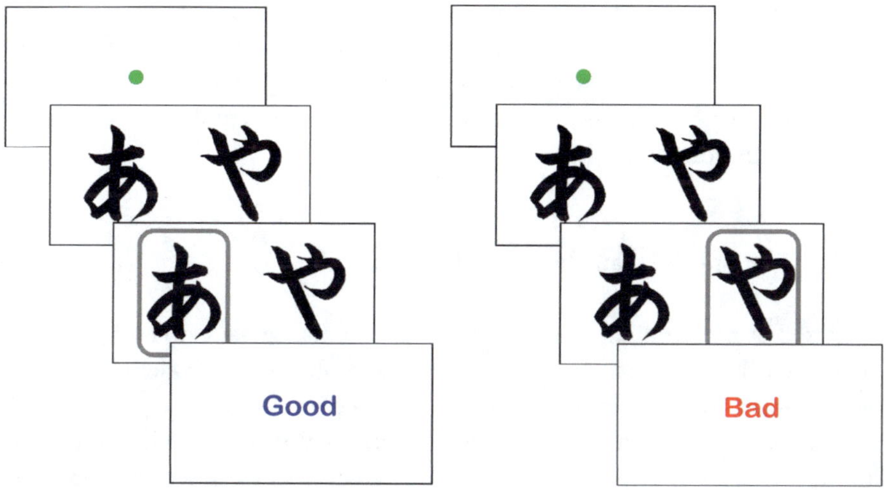

FIGURE 7.1 The flow of a single trial in the test

rewards and penalties received during trials. In the game in question, we have 3 pairs of symbols with different values of probabilities *p*, for example, pair one (0.8 – the probability of receiving a reward when choosing the first symbol, 0.2 – the probability of receiving a reward when choosing the second symbol), pair two (0.7, 0.3) and pair three (0.6, 0.4). The pairs are presented in random order. The easiest pair to learn is the first pair with extremely different reward probabilities, while the hardest pair to learn is the third pair. Because the rewards are given in a probabilistic manner, and because of the complexity of the symbols, knowledge of the hierarchy from most to least rewarded symbols is stored in implicit memory, i.e., the patient has only an intuition that certain symbols are better than others.

In addition, before starting this test, the player is informed that one symbol is better than the others. This symbol is presented, so that the subject can remember this symbol in explicit memory (this symbol is called an instruction). This information is false, and on subsequent trials, it is observed that the player's tendency to abandon the false instruction is in favour of the symbols that are most often rewarded during the game. This modification enables the recognition of the activity of additional brain structures involved in the processing of information provided at the conscious/implicit level.

During the implementation of the test, data is collected on the successive decisions made, the rewards and punishments received in subsequent trials. This information is called behavioral data. Once the test subject has performed the test, an appropriate model can be fitted to the collected data, and by estimating the parameters of a given model, it is possible to characterise the test subject using variables such as sensitivity to punishments and rewards, exploration drive (a tendency to make a risky decision) or the effect of instruction on learning. In the next sections, we will present neural network models and computational models.

3 Biologically Plausible Neural Networks

For the first time in 1941, John von Neumann, a mathematician and physicist, developed the automata theory, in which the brain can be modelled as a network composed of nodes that are individual nerve cells. A description of such a network includes various cell states, the formulation of a rule for processing impulses reaching a specific cell from other cells, and connections between cells (Neumann, 1956). In 1943, the neuropsychiatrist Warren McCulloch and the mathematician Walter Pitts proposed a relatively simple computational model of a single nerve cell and proved that a network composed of such neurons has the computational capabilities of a Turing machine (McCulloch & Pitts, 1943). This means that any computable function can be evaluated using a neural network containing a sufficient number of neurons connected in a proper way (Hornik et al., 1989). Although the theorem asserts that it is feasible, it provides no guidance on how to accomplish the goal. As a result, neural network design is based on trial-and-error. Nowadays, engineers create neural networks with billions of parameters that can do tasks like classification, prediction, categorization, adaptive process control, language processing, and so on (Sarker, 2021).

However, artificial neural networks employed in commercial artificial intelligence (AI) solutions must be distinguished from biologically plausible neural networks. Contrary to the common belief of most computer scientists, the former has nothing to do with biological neural cells. The latter, however, are precise models of the electrical activity of brain cell networks. Commonly used in AI deep networks are a lengthy sequence of alternating matrix multiplication followed by a nonlinear operation (Chollet & Chollet, 2021). The total number of elements in all matrices representing neural network parameters can reach hundreds of billions (*GPT-4*, n.d.).

3.1 *Modelling the Neuron*

Neural networks used as detailed models of brain processes mirror the dynamics of real neurons, brain structures and neural networks in many aspects.

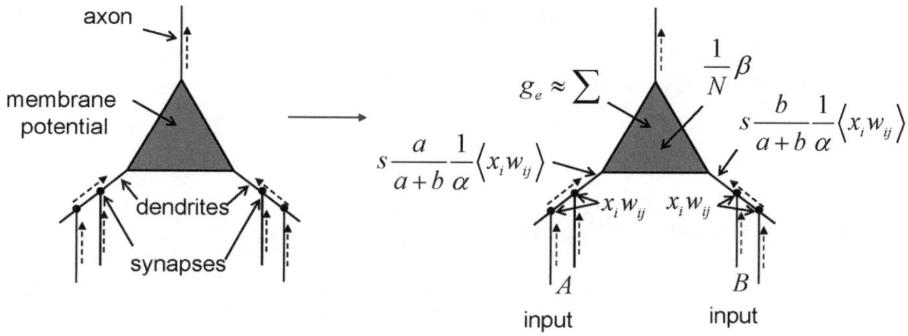

FIGURE 7.2 Description of the neuron model proposed within the Leabra framework (adapted from O'Reilly & Munakata, 2000)

Figure 7.2 depicts a model of a neuron, together with equations describing how its various sections work, proposed by C. O'Reilly (O'Reilly, 1998). From the equations, the action potential appearing at the neuron's output in response to excitations at all inputs of the neuron can be calculated. The model shown is part of Leabra's framework (O'Reilly & Munakata, 2000), which defines the behaviour of artificial neurons to stimuli, how they are connected and the detailed algorithms for changing the synaptic weights that enable the network to learn. In this case, the network learning algorithm is a variation of the so-called Hebb's rule. In the above-mentioned work, a full set of eighteen equations of the Leabra mathematical framework is given, three of which describe the operation of the neurons, three of which relate to Hebb's rule, two of which are used to control the learning rate, and the rest that play a supporting role (e.g., determine the equilibrium state of the neuron).

The main advantage of the Leabra framework is the speed of computer simulation of the neural network. However, this comes at the cost of some simplifications. The neurons described cannot model many of the dynamic behaviours of real neurons. Nowadays, developments in computer technology and methods for simulating dynamic systems make it possible to execute more realistic models by computer. A gold standard in the field are *Spiking Neural Networks* (E. M. Izhikevich, 2003). Izhikevich lists all classes of the behaviour of biological neurons and shows that the neurons he proposes can reproduce the basic properties of their functioning once properly configured (E. M. Izhikevich, 2004). The Izhikevich model of a single neuron reduces to two differential equations:

$$v'(t) = 0.04v(t)^2 + 5v(t) + 140 - v(t) + I(t) \quad (1)$$

$$\upsilon'(t)=a\big(b\upsilon(t)-\upsilon(t)\big) \tag{2}$$

and one conditional rule: whenever

$$\upsilon(\tau)\geq 30$$

for some time instant τ, apply the following substitutions:

$$\upsilon(\tau)\leftarrow c,\ \ u(\tau)\leftarrow u(\tau)+d \tag{3}$$

The membrane potential is v, u stands for the regenerative variable and I denotes the current flowing into the neuron. The parameters a,b,c,d determine the dynamic characteristics of the neuron. The Izhikevich model was developed as a simplification of the four-dimensional Hodgkin-Huxley model to two dimensions (E. Izhikevich, 2006).

3.2 *Modelling the Network of Neurons*

A neural model is a realization of a specific hypothesis about the organisation in the brain of the learning process, starting with how neurons process information, their connection in the neural network layers and the structure of selected brain areas. The design of a biologically plausible neural network model comprises the mathematical model of a single neuron (as shown above) as well as the connections between these neural units. Behavioral data from various neuropsychological tests and brain function data from measurement tools such as *electroencephalgoraf* (EEG), *functional magnetic resonance imaging* (fMRI) or structural *magnetic resonance imaging* (MRI) may be used to determine the connections between brain structures of interest. These modern brain imaging tools and the development of mathematical modelling techniques make it possible to develop increasingly detailed and *biologically realistic* neural network models that describe the operation of specific parts of the brain. A prominent example of a computer framework for building large-scale simulation models of brain functions is NENGO, which focuses mainly on mimicking the cognitive functions of a human brain (Bekolay et al., 2014; DeWolf et al., 2020).

In our studies (Drapała et al., 2021; Drapała & Frydecka, 2022; Frydecka et al., 2016), we choose a medium-scale Leabra model introduced by O'Reilly (O'Reilly & Frank, 2006; O'Reilly & Munakata, 2000), and used by many groups of researchers (Baladron & Hamker, 2015; Gerstner et al., 2014; Humphries et al., 2006; Morrison et al., 2008; Schroll et al., 2014). It consists of layers made up of several groups of neurons. Each group of neurons in a layer contains a

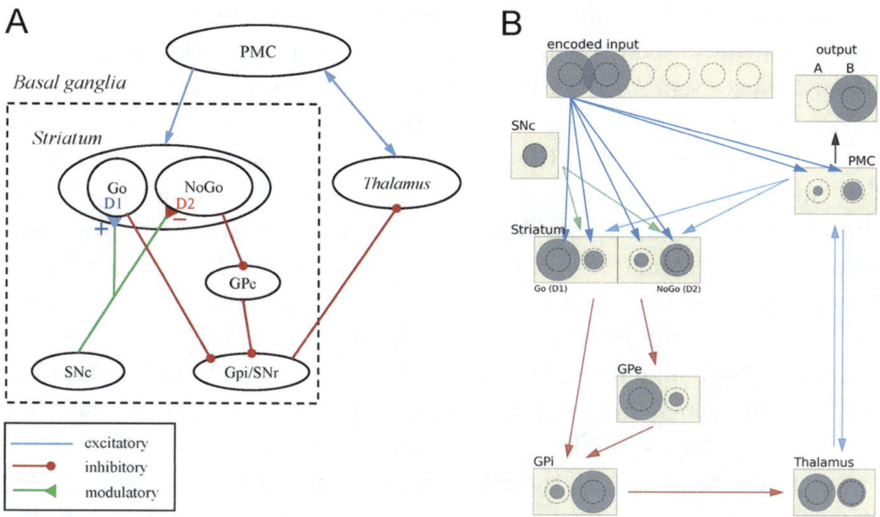

FIGURE 7.3 Conceptual model of *basal ganglia* with corresponding neural model (after: (Samson et al., 2010)). PMC – *premotor cortex,* SNc – *substantia nigra – pars compacta,* GPe – *globus pallidus external segment,* GPi/SNr – *globus pallidus internal segment/substantia nigra – pars reticulata,* Go, NoGo – two distinct types of striatal neurons

number of neurons that transmit signals to other layers, or inhibit the activity of neighbouring neurons within the layer (implementing, for example, the kWTA algorithm). Figure 7.3 panel A shows a diagram of the connections between the layers, while Figure 7.3 panel B illustrates the model in the form of the corresponding neural network (Frank, 2005, 2006).

3.3 *Modelling the Learning Process*

To grasp the essence of the learning process that occurs in a neural network, it is best to follow the flow of signals in the situation of reward and punishment. The Leabra framework specifies two phases of a neural network's learning process: the *minus* phase and *plus* phase. In short, the *minus* phase relies on processing the information about provided options throughout the whole network to determine which option is picked. The *plus* phase begins immediately after the feedback (reward or punishment) is received. The network performs the identical computations as in the previous phase, except for the activity of dopaminergic neurons being affected by the feedback signal. The learning process makes use of the changes in neuron activity in both phases. Let us now get into the details of the *minus* and *plus* phases.

In the *minus* phase, a coded pair of symbols appears at the input of the network. The input layer stimulates the *striatum* (*striatum*) and *premotor cortex* (*premotor cortex*). The differential activity of the premotor cortex neurons is sufficient to decide which symbol should be selected. This will only actually be noticed, once a signal arrives at the premotor cortex from the *thalamus* (*thalamus*). The thalamus thus acts as a decision "gate", accepting or rejecting the decision pre-determined by the premotor cortex. The striatum is made up of two groups of neurons: the *Go group* and the *NoGo group*. *Go* neurons inhibit the activity of the *globus pallidus internal segment/substantia nigra – pars reticulata* (GPi/SNr) complex, which in turn inhibits the activity of the thalamus. Eventually, *Go* neurons unblock the thalamus from inhibiting. Signals from *NoGo neurons* of the striatum additionally have the *globus pallidus external segment* (GPe) bound to the striatum and GPi/SNr by inhibitory connections in their path, which ultimately inhibits thalamic activity. Thus, *Go* and *NoGo* neurons exert opposing effects on the thalamus, stimulating the selection of the appropriate symbol. In the minus phase, dopaminergic neurons located within the grey matter (*substantia nigra – pars compacta,* SNc) show average activity. *Go* neurons receive signals from dopaminergic neurons via D1 dopamine receptors, resulting in additional stimulation, while *NoGo neurons* have D2 receptors, which – in response to the appearance of dopamine – produce an inhibitory effect.

Network learning takes place in the *plus* phase. Generally, the same signals are given to the input of the network and are processed in the same way, but this time the activity of dopaminergic neurons in the SNc changes. If a reward is given for a choice made in the minus phase, the activity increases significantly above average. If a punishment is received, the opposite effect appears – the activity of dopaminergic neurons decreases significantly. Then, the strengths of synaptic connections in selected parts of the network (connections between input, striatum and premotor cortex) change according to the differences in neuronal activity in the minus and plus phases. The process of changing synaptic connection strength is assumed to follow either version of Hebb's rule.

It should be noted that there is an ongoing debate over the biological plausibility of contemporary frameworks that characterise the neural foundation of cognitive abilities of the human brain (Fernández et al., 2021; Jeon & Kim, 2023). The framework presented here is not the only one in the area.

Neural networks fall into the category of high-complexity mathematical models and allow for the representation and validation of precise hypotheses at the brain circuitry level. Understanding cognitive abilities, however, is not the same as reducing everything to neural dynamics. We instead wonder if complex brain dynamics may be conceived as following some simple principles that define the learning process that we directly observe. To put it another way, can we define a simple model that operates at the level of mental functions rather than its low-level neural foundation? Similarly, to understand the behavior of gases under different conditions, we do not need to track individual molecules' motions, but may simplify the model to a few parameters: pressure, volume, and temperature. In the field of computational neuroscience, such simplified models are known as *computational models*.

4 Computational Models

A computational model is a set of equations that directly express the simple rules governing learning and decision-making (Doll et al., 2009; Niv, 2009). The most relevant in the context of probabilistic learning tasks are the so-called *reinforcement learning models* (Sutton & Barto, 2021). We present the *Q-learning* model, which is the most prevalent kind of this class (Samson et al., 2010). Assume there is just one pair of symbols (A, B) in the game to keep equations short. The computational model must account for two stages of brain activity in the subject: decision-making and adaptation/learning.

4.1 *Decision-Making*

According to brain imaging studies (Lee et al., 2012), the neural system of a participant playing a reinforcement learning game represents the expected rewards of selecting an option as the level of activity of a group of neurons linked with this choice throughout the duration of the game (Doll et al., 2009; Samson et al., 2010). The expected rewards for selecting symbols A and B are represented as Q_A and Q_B, respectively. These are real-valued numbers that are often scaled down to fall between 0 and 1. It is reasonable to assume that a subject prefers the symbol associated with the greater Q-value. However, people differ in how much of a difference between two Q-values affects their choices. These individual differences are accounted for by the model parameter T. In other words, T represents the propensity of a person to make risky decisions.

A mathematical model representing decision-making is a formula for $p(A)$, which is the probability that a subject with the T parameter selects the symbol A, given the difference between Q_A and Q_B:

$$p(A) = \frac{1}{1 + e^{-\frac{Q_A - Q_B}{T}}} \tag{4}$$

Obviously,

$$p(A) = 1 - p(A)$$

because the probabilities must add up to 1.

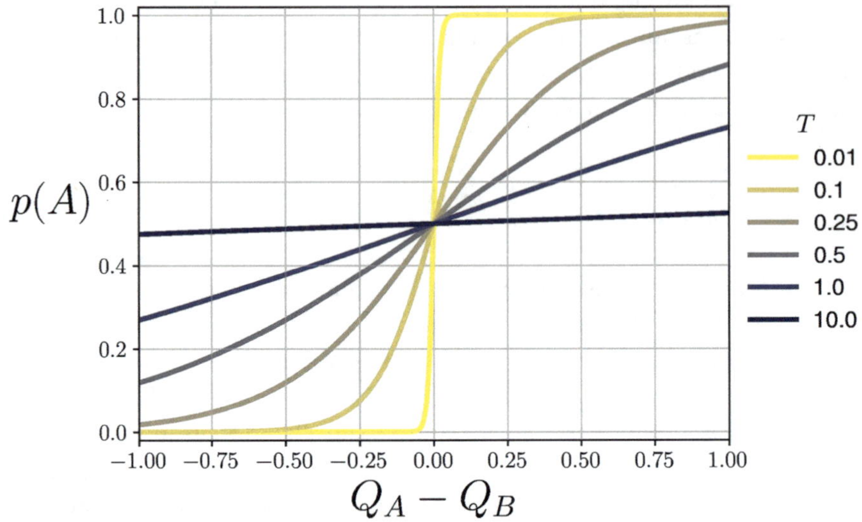

FIGURE 7.4 Graphical interpretation of formula (4)

Figure 7.4 illustrates the key properties of formula (4), known as *sigmoid function*.

If the subject expects a similar reward for both symbols *A* and *B* (the expression $Q_A - Q_B$ takes a value close to zero), then the odds of selecting both symbols are identical. This is exactly what happens at the beginning of the game, when no feedback has yet been provided. We expect a rational player to be less inclined to pick symbol *A* if the expected payoff for doing so is smaller than the expected value for choosing symbol *B* (the expression $Q_A - Q_B$ takes a negative value). However, the rate at which these odds decrease as the advantage of symbol *B* over *A* increases is determined by the magnitude of the parameter *T*. A person with a high *T* value will be more likely to take risks by picking a symbol for which she/he has received fewer rewards in the past. This is reasonable to some extent, since a subject may receive a series of punishments despite selecting the better symbol.

4.2 *Learning*

The game consists of a series of "trials and errors" in which a subject is asked to make a decision and then receives feedback. The variable r will represent the feedback information, with $r = 1$ representing the reward and $r = 0$ representing the punishment. Before the game begins, each *Q*-value is set to 0.5. After completing each trial, the subject can use the feedback information to update the *Q*-values:

$$Q \leftarrow Q + \alpha(r - Q) \tag{5}$$

where

$$\alpha = \begin{cases} \alpha_{\text{Gain}} & \text{for } r = 1 \\ \alpha_{\text{Lose}} & \text{for } r = 0 \end{cases}$$

This equation runs only for the *Q*-value that represents the symbol that was chosen during the current trial. The parameter α is known as the learning rate and it controls the amount of updates taken by *Q*-values. It can take two values: α_{Gain} when the subject receives a reward and α_{Lose} when the subject receives a punishment.

The update formula (5) is easy to follow. When a reward is received, the r–Q is positive and the *Q*-value of a chosen symbol increases. When a punishment is given, r–Q becomes negative, and the *Q*-value decreases.

4.3 *Interpretation of Behavioral Data*

Because of its simplicity, the computational model gives a short summary of behavioral data pertaining to the subject's performance on a probabilistic learning task. The behavioral data contains detailed information on all events that occurred throughout the game's trials. This is a list of symbols that show on the screen, a subject's choices, and the rewards/punishments that follow from those choices.

The governing formulas of the computational model require only three parameters to be determined: α_{Gain}, α_{Lose} and T. This implies that only three numbers are sufficient to characterize every subject. The maximum likelihood method may be used to estimate the values of model parameters using behavioral data, as demonstrated in (Częstochowska et al., 2019). This method condenses a lot of raw data regarding game-related events into three figures that might serve as the foundation for a story: what motivates a subject more, reward or punishment, and what are their propensities to make risky decisions? A complex model like a neural network could not accomplish this. Finally, the parameters of computational models can be correlated with the data from other sources, including genetic data, neuroimaging data parameters (MRI, fMRI), neurotransmitter level data, and others.

5 Relationship between the Computational Model and the Neural Network

The quantity r–Q that appeared in the computational model has an important equivalent in neurology. Mathematically, it is the difference between the actual and predicted value of the reward for a particular decision. At the neurological level, it is closely connected with the activation of dopaminergic neurons in the subcortical nuclei. The decision-making process is concisely described by formula (4), which also masks the complexity of the network in which many neurons collaborate to get a final decision on which symbol to choose. The simplicity of equation (4) emerges from the complexity seen at the neuronal level. Formula (5) similarly captures the complexity of the interactions between the striatum and thalamus in a simple manner.

6 Concluding Remarks

Computational models and neural networks are complementary in that they offer viewpoints on learning from opposite ends of the complexity spectrum. Current neuron models in the literature are so physiologically convincing that even experts may struggle to tell the difference between a measured and simulated signal (Baladron & Hamker, 2015; Gerstner et al., 2014; E. M. Izhikevich, 2004). Biologically plausible neural networks designed to mimic subnetworks of a certain portion of the brain exhibit similar dynamics and can perform comparable tasks. Furthermore, there are known cases of explaining some cognitive disorders in humans by simulating damage in an artificial neural network based on understanding the impact of a given disease on the brain. It is also possible to examine the effects of certain psychotropic drugs on cognitive performance by modelling their effects on the functioning of the neural network (Geana et al., 2022).

However, the goal of computer simulations is to understand the causes and effects of events rather than just mimicking them. The extraction of high-level rules that a complex neural network follows is a challenging process. The problem is that the complexity of an artificial neural network is still too large, maybe comparable to the complexity of specific parts of the brain that are prone to modelling. If the complexity of the model is comparable to that of the original process, we lose the opportunity to grasp the modelled phenomena in simple terms.

In contrast, computational models are far less flexible and can simulate only precisely defined cognitive tasks. They can be provided when researchers have succeeded in grasping the underlying principles of a complex process. We would like to end by quoting a Romanian sculptor Constantin Brâncuși: "*Simplicity is complexity resolved*".

References

3D simulation and evolution. (n.d.). Retrieved June 28, 2024, from http://www.framsticks.com/

Baladron, J., & Hamker, F. H. (2015). A spiking neural network based on the basal ganglia functional anatomy. *Neural Networks: The Official Journal of the International Neural Network Society*, *67*, 1–13.

Bekolay, T., Bergstra, J., Hunsberger, E., Dewolf, T., Stewart, T. C., Rasmussen, D., Choo, X., Voelker, A. R., & Eliasmith, C. (2014). Nengo: A Python tool for building large-scale functional brain models. *Frontiers in Neuroinformatics*, *7*, 48.

Chollet, F., & Chollet, F. (2021). *Deep learning with Python* (2nd ed.). Simon and Schuster.

Częstochowska, J., Duda, M., & Cwojdzińska, K. (2019). Computational investigation of probabilistic learning task with use of machine learning. *Proceedings of 39th International Conference on Information Systems Architecture and Technology–ISAT 2018: Part II*. https://link.springer.com/chapter/10.1007/978-3-319-99996-8_30

DeWolf, T., Jaworski, P., & Eliasmith, C. (2020). Nengo and low-power AI hardware for robust, embedded neurorobotics. *Frontiers in Neurorobotics*, *14*, 568359.

Doll, B. B., Jacobs, W. J., Sanfey, A. G., & Frank, M. J. (2009). Instructional control of reinforcement learning: A behavioral and neurocomputational investigation. *Brain Research*, *1299*, 74–94.

Drapała, J., & Frydecka, D. (2022). Coarse-grained neural network model of the Basal Ganglia to simulate reinforcement learning tasks. *Brain Sciences*, *12*(2). https://doi.org/10.3390/brainsci12020262

Drapała, J., Frydecka, D., & Świątek, J. (2021). A simple-units complex-structure neural network model of the Basal Ganglia to simulate reinforcement learning tasks. *Procedia Computer Science*, *192*, 281–290.

Fernández, J. G., Hortal, E., & Mehrkanoon, S. (2021). Towards biologically plausible learning in neural networks. *2021 IEEE Symposium Series on Computational Intelligence (SSCI)*, 1–8.

Frank, M. J. (2005). Dynamic dopamine modulation in the basal ganglia: A neurocomputational account of cognitive deficits in medicated and nonmedicated Parkinsonism. *Journal of Cognitive Neuroscience*, *17*(1), 51–72.

Frank, M. J. (2006). Hold your horses: A dynamic computational role for the subthalamic nucleus in decision making. *Neural Networks: The Official Journal of the International Neural Network Society*, *19*(8), 1120–1136.

Frydecka, D., Drapala, J., Kłosińska, E., Krefft, M., & Misiak, B. (2016). Computational modeling of reinforcement learning using probabilistic selection task and instructional probabilistic selection task. *European Psychiatry: The Journal of the Association of European Psychiatrists*, *33*(S1), S138–S138.

Geana, A., Barch, D. M., Gold, J. M., Carter, C. S., MacDonald, A. W., 3rd, Ragland, J. D., Silverstein, S. M., & Frank, M. J. (2022). Using computational modeling to capture schizophrenia-specific reinforcement learning differences and their implications on patient classification. *Biological Psychiatry. Cognitive Neuroscience and Neuroimaging*, *7*(10), 1035–1046.

Gerstner, W., Kistler, W. M., Naud, R., & Paninski, L. (2014). *Neuronal dynamics: From single neurons to networks and models of cognition*. Cambridge University Press.

GPT-4. (n.d.). Retrieved June 29, 2024, from https://openai.com/index/gpt-4-research/

Hornik, K., Stinchcombe, M., & White, H. (1989). Multilayer feedforward networks are universal approximators. *Neural Networks: The Official Journal of the International Neural Network Society*, *2*, 359–366.

Humphries, M. D., Stewart, R. D., & Gurney, K. N. (2006). A physiologically plausible model of action selection and oscillatory activity in the Basal Ganglia. *The Journal of Neuroscience: The Official Journal of the Society for Neuroscience*, *26*(50), 12921–12942.

Izhikevich, E. M. (2003). Simple model of spiking neurons. *IEEE Transactions on Neural Networks/a Publication of the IEEE Neural Networks Council*, *14*(6), 1569–1572.

Izhikevich, E. M. (2004). Which model to use for cortical spiking neurons? *IEEE Transactions on Neural Networks/a Publication of the IEEE Neural Networks Council*, *15*(5), 1063–1070.

Izhikevich, E. M. (2006). *Dynamical systems in neuroscience: The geometry of excitability and bursting*. https://doi.org/10.7551/mitpress/2526.001.0001

Jeon, I., & Kim, T. (2023). Distinctive properties of biological neural networks and recent advances in bottom-up approaches toward a better biologically plausible neural network. *Frontiers in Computational Neuroscience*, *17*, 1092185.

Lee, D., Seo, H., & Jung, M. W. (2012). Neural basis of reinforcement learning and decision making. *Annual Review of Neuroscience*, *35*, 287–308.

McCulloch, W. S., & Pitts, W. (1943). A logical calculus of the ideas immanent in nervous activity. *The Bulletin of Mathematical Biophysics*, *5*(4), 115–133.

Morrison, A., Diesmann, M., & Gerstner, W. (2008). Phenomenological models of synaptic plasticity based on spike timing. *Biological Cybernetics*, *98*(6), 459–478.

Neumann, J. V. (1956). Probabilistic logics and the synthesis of reliable organisms from unreliable components. In C. E. Shannon & J. McCarthy (Eds.), *Automata studies (AM-34)* (pp. 43–98). Princeton University Press.

Niv, Y. (2009). Reinforcement learning in the brain. *Journal of Mathematical Psychology*, *53*(3), 139–154.

O'Reilly, R. C. (1998). Six principles for biologically based computational models of cortical cognition. *Trends in Cognitive Sciences*, *2*(11), 455–462.

O'Reilly, R. C., & Frank, M. J. (2006). Making working memory work: A computational model of learning in the prefrontal cortex and basal ganglia. *Neural Computation*, *18*(2), 283–328.

O'Reilly, R. C., & Munakata, Y. (2000). *Computational explorations in cognitive neuroscience: Understanding the mind by simulating the brain.* MIT Press.

Ridley, M. (1994). *The Red Queen: Sex and the evolution of human nature.* Penguin UK.

Samson, R. D., Frank, M. J., & Fellous, J.-M. (2010). Computational models of reinforcement learning: The role of dopamine as a reward signal. *Cognitive Neurodynamics, 4*(2), 91–105.

Sarker, I. H. (2021). Deep learning: A comprehensive overview on techniques, taxonomy, applications and research directions. *SN Computer Science, 2*(6), 420.

Schroll, H., Vitay, J., & Hamker, F. H. (2014). Dysfunctional and compensatory synaptic plasticity in Parkinson's disease. *The European Journal of Neuroscience, 39*(4), 688–702.

Sutton, R. S., & Barto, A. G. (2021, December 1). *Reinforcement learning.* The MIT Press, Massachusetts Institute of Technology.

CHAPTER 8

Art

A Particularly Sophisticated Simplicity

Łukasz Huculak

Abstract

The perceived simplicity of our visual sensations often masks a profound complexity, not necessarily evident in the quantity of connected elements but in the quality of their interconnections. By analogy, what appears visually complex may be the result of certain primitivism: a simple relation aimed at recreating the illusion of complexity, constructed through mechanical multiplications. This chapter explores simplicity and complexity in terms of their interconnectedness within the elements of a picture, where simplicity subordinates details to the whole and complexity disregards the whole for the sake of details. The constant circulation of simplicity and complexity in the process of evolution of the visual arts is emphasized. Taking into consideration art history's progression from simplicity to complexity, (the visual conventionality of Lascaux Symbolism to the photographic literalness of 20th-century Hyperrealism, the transition from Doric to Corinthian styles, and from Renaissance to Baroque), a tendency emerges to equate simplicity with intellectual sophistication while complexity with perceived execution difficulty. Drawing on Immanuel Kant's theses on the power of judgment and Giorgio Vasari's painterly idealism, the text investigates how simplicity unifies and omits detail, while complexity breaks down as differences emerge. Discussing the crisis of form, the text also explores how complexity arises from the dissection of visual elements, while simplicity emerges from bold attempts at synthesis and reduction. Examples such as Mannerism (a breakdown of the whole into its details), collage (unifying accidental details), informality (the multiplying of details as a whole), and academicism (a peculiar "excess of footnotes") illustrate different facets of complexity, while methods like *dominanta* (subordinating all elements to a highlighted area of presentation) and harmony (the favoring of an analogy), exemplified by Johannes Vermeer and Georges Seurat, demonstrate approaches to unification. Ultimately, complexity is not merely about detailing but about the relationships between details, while simplicity arises from unsophisticated relationships, resulting in an apparent complexity in the midst of incomprehensible excess.

Keywords

primitivism – simplicity – complication – complexity – whole – fragment – Vasari – Kant – Vermeer

1

Simplicity and complexity – with an opposition formulated in this manner, it is probably difficult to find a discipline as fortunate as the visual arts. It seems that by applying a universal pattern of evolution from simple forms to ever more complex ones onto the chronology of our artistic operations, we would receive a precise model of its historical development across all schools. Fully realistic visualizations come after the strict simplicity of symbolic pictograms, the flourishing Corinthian style comes after the ascetic Doric style, the fragmented Gothic style comes after the cohesive Romanism, and after the classicism of Renaissance – comes the elaborate Baroque. However, after complex analytical Cubism comes simple synthetic Cubism.

That last example shows that the dialectic of simple–complex sets up the primeval axis of struggles pertaining to aesthetic preferences; it cannot entirely be regarded as a vector programming their development. The periods in which the exquisite, elaborate, and structurally fragmented forms reigned are often only an introduction to reaching for extreme simplicity; they mobilize bold reductions. Contrary to what might be inferred from Darwinist theory, complexity in the visual arts is not dialectically privileged, and in particular the art of the 20th century facilitates multiple examples that prove otherwise.

The persistence of modernist concepts, which explicitly favors primitivism, or the enduring suspicion of the historians towards an elaborate yet eclectic (complex) post-modernism, suggests that the interdependency of both categories is more ambiguous. In the artistic practice of particular authors, it happens that they are equivalent sources and they become not the aim of the activity but one of its tools. The universal mechanism of the alternation of simplicity with complexity and vice versa descends in this way from the level of historical necessity and indicates directions of stylistic ideology to the level of individual choices.

What determines these choices? Is simplification an effect of the natural limitations of human imagination, unable to fully see the potential of the diversity of forms? Or maybe complexity is only a momentary victory of greediness, a vaccination reinforcing noble discretion? It turns out that contrary to intuitive preferences, which locate an admirable executive difficulty in respect, the aesthetics with intellectual sophistication rather join simplicity, whereas complexity is more often tied to emotional immoderation: ostentation, sensual insatiability, extravagance and splendor. Following the slogan "less is more", what is complex is actually unripe, and reduction would be a peak stage of development. Its cumulativeness would consist of the rise of awareness of what should be rejected and not what should be added.

To substantiate these beliefs, the discovery of the "judging power" was undoubtedly important. The remarkably intricate art of Mannerism, which Immanuel Kant viewed as highly susceptible to sensual temptations, was a directly criticised due to that particular excess of form. The judging power crowning the entire cognitive will, provides the aesthetic judgment with transcendental grounds. Kant's aesthetics aiming at putting sensual experience in brackets and turning art into an object for purely mental reflection, put sensually tempting things in contradiction with sophisticated restraint. However, does allure always have to refuse access to speculation and should intellectual attractiveness be compensated with sensual bareness?

One of the biggest dangers of simplicity is that for the good of the totality, it ignores details that do not fit – it cuts off uneven edges, omits places in which separate sets overlap, losing its expressiveness due to the irregularity of borders. It is an action focused on the ordering experience, basically intellectual and yet idealistic. Just like the division of all artistic schools into those that pertain to the senses and those that appeal to the intellect seems to be orderly but is at the same time arbitrary.

When Duchamp was elevating Conceptualism, he placed both of these areas in distinct opposition having had the many centuries of traditional Platonism behind him. It minimized the importance of senses to the benefit of the intellect, and dualism sanctified with academic theory: paint stroke – line and intuition – speculation. Those were expressed by the rivalry of painterly Venice with the sketching of Florence, the fleshy "Rubensists" with rational "Poussenists" or reflective geometrical abstraction with unbridled *taschisme*.

The visual richness and mimetics refer easily only to the senses whereas we regard the lucid Classicism and aesthetic Minimalism as intellectual, however, the complexity of exalted Mannerism or the richness of the persuasive Baroque essentially relied on a sophisticated aesthetic theory which was the effect of the dynamic development of optics, Cartesianism and high self-awareness as well as relativism (Stoichita, 1997).

One of the gratest dangers of complexity lies in its ability to discernsubsequent nuances, multiply countless divisions, and indicate further differences within similarities and analogies within oppositions. However, it often pays for its insightfulness with the clarity of the entirety. Essentially, speculativeness (over-intellectualizing) should be responsible for art following the path of complexity, and the main culprit on trial should be Giorgio Vasari – its first theoretician. Vasari's theory aimed at elevating art from the level of crafts to the level of subordinate *science ratio*. It has regarded an artist's work as a materialization of ideas, which meant not only its transcendental sources but also the optimization of composition and color-related solutions. A painting was supposed to be a "machine", to which nothing can be added or subtracted without harming it. Although there was a chance of reaching ideal simplicity, it ended up exactly the opposite – with complication and entanglement.

The search for the optimum state is usually done in the cycle thesis – anti-thesis often delaying the conclusion *ad finitum*, burying every synthesis in the endlessness of potential solutions. Mannerism also left countless *non finito* abandoned along the way. A work of art, to which nothing can be added or taken away without damage is by definition too prone to subjectivism. An artist, polluted with the faith that there's a possibility to bring such a being to life, against the known truth that "the devil is in the details" is often condemned to Sisyphean torments.

That is why Vasari had to look for further guidelines and he found them in the opinion of the majority: instead of endlessly searching for the one, ideal solution, he recommended compiling them out of the Great Masters' achievements regarded as eminent: Rafael's head, Michael Angelo's musculature, Titian's colors (Vasari, 2003). As a result of that democratization, the extreme complication of forms was further compounded by their eclectic recyclability. New solutions could only be the sum of the previous ones and the complexity – putting things together, that connect many separate fragments until that moment – became an academic standard.

Is that the reason why the periods of complexity should be related to the escalation of the artistic crisis? Should the domination of simplicity that followed after them be associated with overcoming doubts and progress? Essentially, eclectics should be an expression of some lack of faith in one's own abilities, looking for shelter under the wings of authorities honored by Vasari in his immortal *Lives of Painters*. That is however not the end of the academician's faults.

In favor of additional complexity was the crucial role that Vasari in his theory ascribed to the line. In drawing – the intellectual analysis of a form – the key method of visualization is setting the boundaries – the division. When the "naturalist" Leonardo da Vinci softened the line, which did not exist in nature, with his painterly *sfumato*, and Cézanne (not having other ambitions than a maximum approximation to nature) rejected it completely, the effect was synthesis and simplification based on the gradual blurring of unambiguous shapes. Through glazes, Leonardo achieved the rejection of strict borders, while Cezanne thanks to geometry achieved, the reduction of all the elements of painting to a common denominator. Those are tools that allow one to organize a flat surface not according to analytical variety but according to synthetic analogy.

That leads to the conclusion that what essentially differentiates simplicity from complexity in the visual arts is not the opposition: senses – intellect, but the favoring synthesis or analysis as a compositional rule, which determines the paintings' internal relations between the detail and the whole.

Loose relations between the detail and the whole characterize not only Mannerist or Baroque paintings, entering us into perceptional confusion by joining groundlessly empty plains with the excess of detail. In particular, Dadaism was perfect in juxtaposing fragments primarily devoid of aim with its model product – collage, but also with *informel*, in which the whole is often a replication of a fragment.

In all these examples carrying out a synthesis is inhibited: the entirety of an artifact is about the dominance of a detail, excess of details or the lack of relationship between them that disorganizes the whole. That fact unveils the structural similarity of seemingly extremely distant visual schools: realism focused on narration and abstract expression, which are not similar in acting with relationships but in acting with the mass. In the first instance, that is the multitude of descriptive and literal motives and in the second one – the multiplication of varicolored splashes.

A domination of detail additionally assumes a composition-related openness and dynamic, making the process of perception disordered, because the lack of a dominant factor forced the viewer to a constant shifting of the view field. The eye frantically jumps from one detail to another, either attracted by the intensity of contrasts or by searching for mutual relationships, with difficulty keeping up with the focus (Arasse, 2013).

Therefore, both realism and abstract expression are bound to nature, which similarly to the object with prolificacy and entropy against any inclinations of the order setting hierarchizing or reduction. Meanwhile, a painting, which is hardly a fragment of reality, but its section (often subjective), at the same time has to constitute an inbred whole. The simplest method of merging its constituent elements is subordinating all of them to one of them – a dominant. The simplicity of such constructs is however too obvious and foreseeable, often more boring than the primary chaos. Less literal ways have to be used, a system that under the pretence of complexity would operate with the infallibility of simplicity, something which in a way is unclear for the recipient would make the synthetic clarity resistant to analysis. The paintings of Johannes Vermeer excel at using such a mechanism, they are one and the other: chaotic in close-up, but orderly from the distance. Particular elements when subjected to scrutiny turn out to be unrecognizable, amorphously painterly, whereas synthesized from a distance they reveal themselves with absolute clarity (Wheelock, 1995).

That is one of the examples when we do not say appreciatively: look at the way it's done! But instead, we ask astounded: how is it done? Abstract stains, which so strongly legitimize the realism of presentation despite its indirectness, show the "phenomenon in which the same painting is not showing the same thing from up close as from a distance".[1]

The divisionist paintings of Georges Seurat act in a similar, although more "mechanical" way. From up close they show a shapeless, chromatic magma but from a distance, they create clear and recognizable forms.

The relationship between parts and the whole in Vermeer's paintings is strict although not obvious. If we want to understand what can be seen in them, we often move closer trusting that we will see better if we put together a whole out of its parts, if we examine it piece by piece.

In the case of Vermeer, we come across a surprise: from up close his paintings turn out even more unrecognizable, we only see another dot describing another dot, but treated synthetically the areas still look as if we were watching them from a distance, they impact by an indivisible whole. His representations reach both the beginning of the painting process – putting together singular, meaningless brush strokes into a spatial whole, and the beginning of interpretation – the moment in which the painting is still plainly visual, a phenomenon devoid of linguistic references to reality (Marin, 2002).

Thanks to this observation it does not lose itself in reading details which inevitably becomes a reading out of senses. The viewer avoids another potential complication: contextualization. However if one had to choose between Vermeer and let's say Henryk Siemiradzki, the statistical majority would choose the paintings of the latter. This is most likely mainly for his story-telling erudition.

If we want to see Siemiradzki's painting we are most of all forced to recognize and then name and think of the thousands of things arranged around it, which – along with what Ernst Panofsky observed – causes us to no longer see the shapes nor the things but the story they are telling and even more – all the digressions implied by the story. What Siemiradzki extends endlessly with delight by adding further details, Vermeer closes in two precise sentences and although they do not explain everything they are intriguing at least with this understatement. That would be the difference between the condensate of poetry and the extravagance of fiction.

Superficial simplicity may then hide an internal complexity, whereas sophistication could be merely an unimaginative multiplication. What seems simple from afar, up close may turn out to be an exceedingly complicated machine, but also what is complicated from a distance on closer inspection might turn out to be just a scenography. The source of these phenomena is the not-so-obvious relations connecting the artifact's particular pieces into a whole and for their discovery, it is necessary to employ an essential system of reference – the distance which we place ourselves towards the observed object.

The complication of these seemingly simple sets takes place not only through working out of particular elements but the subliming of the relations that connect them, which determine the course of boundaries between the objects, the dynamics of gradients and the clarity of lines. They might be sometimes gentle and then sharp, sometimes regular and then serrated, or they complicate themselves in many different ways. Indicating differences and finding relationships, disconnecting or connecting, but mostly: the dynamics of those changes, how softness shades into hardness, shadow into light and one texture turns into another – that is the secret weapon of a painting.

From the author's point of view, it is fascinating how relatively small shifts in accents, the extraction of one relation or hiding another, completely change the visual structure of the entire representation: how thin is the line differentiating an exceptional effect from a trivial one? Sometimes only one additional point is enough for the entire work to be "complete", to gain character. Other times such an inconspicuous addition throws back an almost finished artifact into an illegible chaos.

The exterior simplicity of visual sensations might be apparent, while essentially it concealsa highly complex mechanism This complexity does not arise from the sheer number of connected elements, but rather from the quality of those connections – the precisely established relationship between a few, selected elements. Analogically, a visible complication might be an effect of essentially banal and simple primitivism that constructs an illusion of complexity through mechanical multiplication. Amongst others, this is proven by the fact that the effective repetition of ingeniously simple things often is a more demanding task than copying a chaos in which precision loses meaning to the benefit of the mass.

A few seemingly uncomplicated parameters might work more intensely than an elaborate system, entering into a less obvious relationship, and at the same time creating a more infallible mechanism.

Although this rule is not necessarily applied to the term "category" in aesthetics, I would distinguish between two kinds of complication (or simplicity): a superficial and an evident – primary and deep. The second one which does not reveal itself at first sight, it exists in an artifact and makes it work. If we assume that things externally complicated may be supported by the structural simplicity, whereas lucid ones are often complex, what is more, that their exterior simplicity is often a result of their internal complexity, I might risk differentiating the *quantitative complexity* (which – unfortunately – complicates matters) right next to its sisterly *qualitative complexity*. That would consist of refining relations between relatively few elements, arising not from multiplicity but from disintegration and the variety of differences – change of line directions and divisions in the reference of compositions, or changes of shades in terms of color – but out of the way that a limited number of differences is composed and mutually matched.

I would look for the reasons of that quality-quantity ambiguousness hidden in reality not less than in people's artifacts: on the one hand, in the features of its structure that are composed both out of details and wholes and, on the other hand, in the particularity of our senses in which observations are not always preceded by conclusions but sometimes exactly the opposite – they depend on them.

The simplicity or complexity of paintings is not solely an issue of their immanent features, but also a subjectivism in observing the differences and two basic ways we acquire impulses: analysis and synthesis. The relation of the distinction "simple – complex" with the senses, on the one hand, and intellectual evaluation on the other, in a wider context may even refer to the relationships of complexity with chaos or simplicity with order. The visual arts would then constitute an attempt to refer to the issue of the codependency of both categories: an ordered whole and complex fragmentation.

The first one, as it appears as simplicity, would hide an internal complexity; the second one would owe the apparent complication to the pressure put on a fragment, obscuring the recognition of the whole. When we soak up the whole we often don't realize their detailed complication and vice versa: while we are lost in the maze of particular parts we see their primary role as richness and complexity. Many elements, although they objectively exist in the painting, keep them "invisible" both due to the crowding of sensations and the limitations of our perceptive apparatus.

Visual simplicity is often an effect of adequate coding of complexity in which the set of simplexes, independently of the number of its components ensures adequate clarity of the whole thanks to a skillful alignment of the properties of the artifact with the rules by which the receiver's perceptive apparatus works. In reference to time, which for spatial visual arts is a certain challenge, it provides a dual effect: complexity extends the time of perception of the artifact, and simplicity provides immediate action, thereby eliminating the temporal extension of the painting in the meaning in which it exists in music.

What is fascinating in Vermeer's paintings is that in their case we do not need time to get to know the whole but we need that time to soak in the unpredictability of spatial relationships, which connect the background with the foreground. Simple paintings – with their visual clarity are much more resistant to our gaze than visual complication, a mechanism devoid of synergy in which the multiplied elements actually act independently without establishing a mutual relationship. Due to that, they require that perception to be extended in time (Didi-Hubermann, 2005).

Nevertheless, it seems that finally, complexity is only a means, a tool to achieve simplicity or rather – to achieve an appropriate complication of that simplicity. This constant circulation of entanglement and simplification proves that overloading one with visual sensations is essential not only to accustom the senses of a recipient to simplicity but mostly to enable the artist to retrieve appropriately not obvious relationships from the maze of sensations. Formal entanglement, which is often an effect of the victory of relativism – the strong conviction of the impossibility of selecting just one, optimal version, has an essential methodological meaning for art. An unrefined extravagance would have a revelatory influence, fertilizing the aesthetics with a dose of new forms, combinations and unexpected effects, unachievable with more "calculated" and rational methods.

Note

1 Didi-Huberman, 2011, p. 157.

References

Arasse, D. (2012). *Detal. Historia malarstwa w zbliżeniu*. Kraków: Dodo Editors.

Arasse, D. (2013). *Take a Closer Look* (A. Waters, Trans.). Princeton University Press.

Didi-Huberman, G. (2005). *Confronting images: Questioning the ends of a certain history of art*. Pennsylvania State University Press.

Didi-Huberman, G. (2011). *Przed obrazem*. Gdańsk: Wydawnictwo słowo/obraz terytoria.

Marin, L. (2002). *On representation* (C. Porter, Trans.). Stanford University Press.

Marin, L. (2011). *O przedstawieniu*. Gdańsk: Wydawnictwo słowo/obraz terytoria.

Stoichita, V. (2011). *Ustanowienie obrazu*. Gdańsk: Wydawnictwo słowo/obraz terytoria.

Stoichita, V. (2015). *The Self-aware image: An insight into early modern meta-painting*. Cambridge studies in new art history and criticism. Cambridge University Press.

Vasari, G. (1988). *Żywoty najsławniejszych malarzy, rzeźbiarzy i architektów*. Warszawa: PWN.

Vasari G. (2003). *Lives of the artists*. Penguin Classics.

Wheelock, A. K. Jr. (1994). *Jan Vermeer*. Warszawa: Artes.

CHAPTER 9

The Complexity of a Straight Line

Jakub Jernajczyk

Abstract

Starting from some reflections on a potentially simple geometric object such as a straight line, this article seeks to show how complex the problems are that we become entangled in when analysing its structure. If we assume that a straight line is made up of points, then it is not obvious how it can be made up of these points, or how we can extract individual points from it. Further considerations will lead us to distinguish between the concepts of potential and actual infinity, allowing us to draw attention to the significant gap between the uncountable complexity of mathematical spaces and the presumed structure of physical reality. As a basis, the article assumes certain common-sense reflections on the issue at hand, relying mainly on visual intuitions and avoiding the formal language of science.

Keywords

dimensionless point – straight line – dichotomous division – cutting – actual infinity – discrete structures

1 Visual Intuitions

One could point to a straight line as a visual equivalent of the concept of simplicity. The straightness of a straight line is, however, not as simple as it seems at first. The straight lines familiar from a checkered notebook, drawn according to the rules of Euclidean geometry, are only a tiny fraction of the set of all possible mathematical straight lines. For example, a line drawn in chalk on a road running along the equator – assuming that such a road could exist – would turn out to be a circle. For there are many different non-Euclidean geometries and lines drawn according to their rules most of us would call curves. It turns out that the simplicity and straightness of a straight line, understood directly,

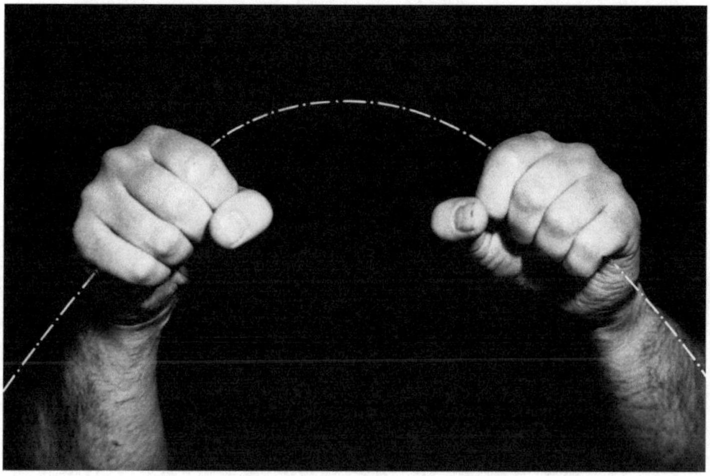

FIGURE 9.1 W. Gołuch, *Double parabole*, 1979

quickly becomes a troublesome concept. Perhaps less difficulty is encountered when we consider the issue of its complexity.

Let's start with the well-known assumption that a straight line is composed of points. Is it somehow testable since "no one has ever seen or touched a point" (Russell 2009, p. 91)? What a straight line and a point are? In what sense do these objects exist? In what way would a straight line be composed of points? These are the kinds of questions that will be the starting point for the considerations in this essay. We will consider mathematical objects and terms but in a rather particular way. For our theoretical toolkit will not be formal definitions and abstract symbols but intuition and visual imagination. So our deliberations will be limited to what can be derived from a common-sense analysis of visual phenomena, both those perceived by the senses and those imagined. The proposed approach is deeply embedded in the tradition of European thought. Aristotle pointed out the crucial importance of visual sensations in the process of cognition in the very first words of Metaphysics (2016, p. 980 a). Summing up the ancient philosophy, Sextus Empiricus explained that what we come to know is either the effect of direct observation or is derived from this observation as a result of inference (2018, III 40). At the end of the Middle Ages, Nicholas of Cusa emphasized that "the power of the mind [...] cannot be stimulated except by the intermediacy of perceptual images" (1996, p. 77). Similarly, Descartes claimed that "sight is the noblest and most comprehensive of the senses" (1985, p. 81). In more recent times, Rudolf Arnheim strongly emphasized the cognitive role of perception and visual imagination (1969).

2 Construction of a Straight Line

We assumed that a straight line consists of points. But in what sense would it be composed of these points, since, according to Euclid's famous definition, "a point is that which has no parts" (Euclid, p. XVIII)? In other words, a point has neither length nor width or depth; a point has a zero dimension. So how can a straight line, which has a length (although it has no width), be composed of dimensionless points? It would be like we wanted to get "something" by adding up "nothing" (Nicolas of Cusa 2000, p. 9), while it is well known that when we add any amount of zeros, we always get a zero. Therefore by connecting points of zero dimension, we cannot receive even the shortest section, and all the more an infinitely extending straight line.

Sextus Empiricus based his argumentation against the geometers on similar doubts (2009, III 34-36). If we assume that a straight line is composed of points lying in a row, these points will be or will not be touching one another. If the points do not touch each other, there are gaps between them, so this arrangement does not form a coherent line. If the points touch each other, there are again two possibilities: they can touch each other in their entirety or with their parts. However, if points touch with their parts, it means that they do have parts, and therefore they are not indivisible, which directly contradicts the assumed definition of a point. Whereas if points touch as a whole, they must be contained within each other, occupying the same place, and as a result, they create a single point, not a straight line. Therefore, regardless of which path our thought would follow, according to Sextus, the conclusion is one: we cannot build a straight line out of dimensionless points.

Ancient scholars also considered another way of forming a straight line: it would be created by a flowing point. However, this explanation was also entangled in logical contradictions, which were pointed out by Sextus Empiricus (2007, III 31-33). Moreover, such a dynamic approach spoke more about how the line can be drawn than its internal structure.

3 Divisions of a Straight Line

If we cannot easily build a straight line out of dimensionless points, let's assume the opposite direction and look inside the structure of the existing

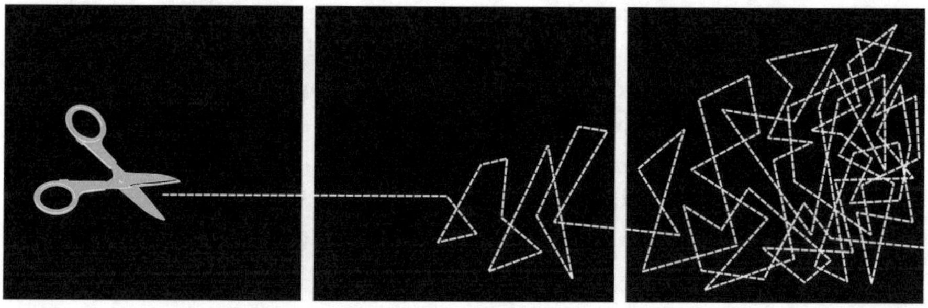

FIGURE 9.2 W. Gołuch, *Cut out that line*, 2011 (parts 0, 9, 10)

straight line, trying to reach the single points. This approach seems even more legitimate since it is easier to imagine a straight line than a dimensionless point. Although the perfect mathematical straight line is never observed in nature, numerous natural phenomena – plant stems or rays of light – as well as products of a human hand – stretched strings or folded sheets of paper – shape up in our minds the basic intuition of line (Davis & Hersh 1981, pp. 126, 158).

So with a straight line in our imagination, let's try to penetrate its structure. The first idea that may come to our mind is separating a section (fragment of a straight line) and then its consequent division. Perhaps by dividing the section into smaller and smaller parts, we will manage to reach a single point. The geometric method of symmetrical divisions of line sections was called by Greek mathematicians *dichotomy* (Euclid, p. 10). Sections obtained as a result of such a division can be subjected to further dichotomous divisions, and this process may be repeated any number of times.

On the basis of infinite dichotomous divisions, Zeno of Elea built his famous argument against the motion, which we know from a short mention by Aristotle. It says that motion does not exist because "…that which is in locomotion must arrive at the half-way stage before it arrives at the goal" (Aristotle 1930, 239 b). Due to the imprecise message, this argument may be interpreted in several ways. The most common version seems to be the following (Kirk & Raven 1975, p. 293):

> It is impossible to traverse the stadium; because before you reach the far end you must first reach the half-way point; before you reach the half-way point you must reach the point half way to it;and so on *ad infinitum*.

We will not devote much time here to rebutting Zeno's argumentation.[1] Let us only mention the solution by Aristotle, who noticed that when Zeno assumed endless divisibility of the path, he did not allow for an analogous – infinitely divisible – treatment of the time in which this movement takes place (Aristotle

1930, p. 233 a). If this unjustified assumption were removed, the surprising conclusions about motion, or rather the lack of it, could not arise.

Returning to the essence of our considerations, one can see that we cannot reach single points of a straight line with the help of dichotomous divisions. Geometrical divisions of any short section always give us sections twice as short but non-zero length. By repeating these divisions even indefinitely, we would never reach a dimensionless, indivisible point.

4 Indicating Points

Many mathematicians, however, are convinced that dimensionless points on a straight line exist, although, as we have already seen, they "are not to be reached by successive divisions" (Russell 1917, p. 83). There must be then a different way to reach them. Would it be possible to indicate them somehow? But how can one indicate something that has no dimension or, more precisely, has zero dimension? It definitely cannot be done with any material indicator because such an indicator, however narrow it may be, will always have some width, infinitely exceeding the zero width of a dimensionless point. The width of the narrowest pointer will always cover a huge, in fact infinite, number of different points. Therefore, we could not consider such an indication to be precise in any case.

There is, however, an indication method that is insensitive to the dimension of an indicator. The method is: cutting a straight line at a certain point. When we cut through any physical or imaginary object, we split it into two parts, which remain on both sides of the blade. There is no part of the cut object at the cutting point – the cutting itself does not remove anything from that object. It makes the dimensionless indication we were after (Jernajczyk 2015, p. 275). Thus, this method allows us to indicate any dimensionless points on a straight line.

For mathematicians, the infinite line provides a model of the set of real numbers in which points correspond to rational and irrational numbers. Based on the visual intuition of cutting a line, Richard Dedekind developed a formal method of construction of real numbers (1872). Although this method has an abstract form, a trace of geometrical references is still present in its name: *Dedekind cuts*.

5 The Universum of Points

Accepting the point structure of a straight line requires us to first accept the so-called *actual infinity*; that is, the kind of infinity that actually exists and is

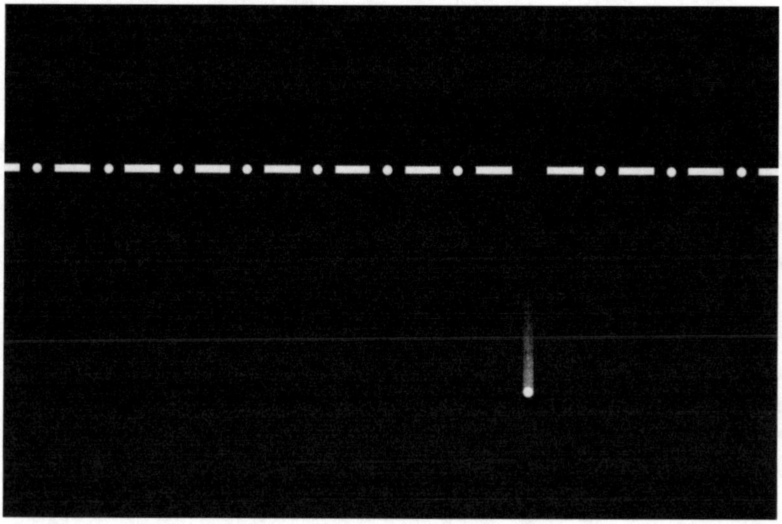

FIGURE 9.3 W. Gołuch, *Falling dot*, 1979

given, as it were, at once, as opposed to *potential infinity*, which is only a possibility, a goal to which one can constantly pursue. Aristotle, who distinguished and defined these two kinds of infinity rejected the existence of the actual infinity right away (1930, pp. 206 a–207 b). Many centuries had to pass before one dared to challenge this categorical decision of the Philosopher.[2]

The paradox conclusions to which the infinity deliberations led did not make that easier. In the fifth century, Proclus Diadochus wrote about a circle divided by its diameter into two semicircles. If we assumed an infinite number of different diameters (plotted at different angles), we would receive twice the number of circle slices. But how can there be something twice as large as infinity? It cannot since actual infinity is unacceptable (Proclus 1992, p. 158).

Twelve centuries later, Galileo Galilei considered two lines of different lengths and, assuming that each is built of an infinite number of points, asked whether the infinity of the longer one is greater than the infinity of the shorter one. Next, he looked at the natural numbers and noticed that each of them can be assigned its square (1–1, 2–4, 3–9, 4–16, etc.). It turns out then that there must be as many natural numbers (infinitely many) as their squares, although the second ones are, after all, the part of the first ones. Finally, Galileo concluded that it is impossible to compare these sets because "the attributes 'equal', 'greater', and 'less', are not applicable to infinite, but only to finite, quantities" (Galilei 1638/1914, pp. 32–33).

What seemed a paradoxical conclusion to the scholars of antiquity and Galileo became, to Bernard Bolzano, the mathematician of the nineteenth century, a positive feature that distinguishes infinite sets (1950, p. 96):

> When two sets are both infinite, they can stand in such a relation to one another that: (1) it is possible to couple each member of the first set with some member of the second [...]; while at the same time, (2) one of the two sets can comprise the other as a mere part of itself.

Based on Bolzano's observation, Richard Dedekind formulated the contemporary definition of an infinite set: "A system S is said to be *infinite* when it is similar to a proper part of itself" (Dedekind 1888, p. 31). In other words, a set is infinite if there is a one-to-one correspondence (*bijection*) between it and some part of it (a subset), which means that we can pair the elements of these sets. In that way, the gates to the actual mathematical infinity became wide open.

Fundamental discoveries in the characterization of infinite sets were made by Georg Cantor. He proved one of the astonishing properties of the set of real numbers: there are "as many" elements in its arbitrarily small interval as in the whole set.[3] Consequently, he precisely justified what had already been sensed before, that even the shortest section, like the entire straight line, consists of an uncountable number of points. Cantor also proved that any section has "the same" number of points as the square and cube built on it. Finally, an arbitrarily small section contains "as many" points as a space with any number of dimensions (Cantor 1877).

6 Discrete Complexity

A continuum of dimensionless points is, of course, not the only possible variant of mathematical space. There are many topological models postulating its various structures (see Gruszczyński 2016). Understanding them, however, requires advanced formal notation, which goes far beyond the framework of visual reasoning adopted here. But we can easily imagine a model of discontinuous space in which there are the shortest indivisible sections and the smallest fields and volumes of space. Concepts of the discrete structure of the physical world appeared as early as antiquity. The most famous was undoubtedly the atomistic theory of matter developed by Leukipos and Democritus of Abdera. However, also in earlier Indian philosophy and later in medieval Arabic philosophy, the existence of atoms was postulated (Jernajczyk 2016, p. 97). And not only atoms of matter but also atoms of time and space.

The concepts of the discrete structure of physical reality returned to the mainstream of science after many centuries. The *loop quantum gravity* is a contemporary theory that assumes the discontinuity of physical reality. According to that theory, like matter, which is made of elementary particles, space is also not infinitely divisible but has the smallest lengths, areas, and volumes, while time is made up of elementary, discrete events. Like the apparent continuity and smoothness of matter, the continuity and smoothness of time and space is only an illusion (Smolin, 2001, pp. 62, 95). Today, it is much easier to accept such a concept since we interact daily with illusions that operate on similar principles. For example, we observe the apparent smoothness and fluency of static and moving digital images, although there are no doubts that their structure is discontinuous (see Jernajczyk 2013). For it is enough that these smallest, indivisible, temporal, and spatial elements are below the threshold of our perception so that what has a discrete structure can create an impression of continuity.

The adoption of a discrete structure may raise doubts about its potential complexity. Could such a limited model cope with the variety of phenomena we observe in nature? Already the ancient apologists of atomism pointed in this context to the alphabet, which allows us to write any conceived content using a small set of characters. All the more, an innumerable set of atoms can produce the perceived richness of nature (see Lucretius 2001, p. 25). In the 20th century, Jorge Luis Borges described a library that would contain all possible to write books in a fixed format. The staggering number of volumes in this hypothetical collection, resulting from the combinatorial calculation,[4] allowed the author to call this library the universe (Borges 1964). Even more complex is a high-resolution digital image composed of2 073 600 pixels (1920x1080), while each pixel can display 16 777 216 different colors (in the True Color standard). The number of all possible images created on such a matrix equals the unimaginable value of $16\ 777\ 216^{2\ 073\ 600}$ (Jernajczyk 2018, p. 216). Thus, it is clearly visible that even limited discrete structures make it possible to achieve complexity far beyond the limits of our imagination.

7 The Limits of an Image

We started our considerations with the most basic objects – a straight line and a point – and came to uncountably infinite sets. From an insight into the structure of a single section, in a few steps, we came to questions about the complexity of the entire universe. However, it should be remembered that at a certain stage, the possibilities of pictorially grasping abstract issues exhaust

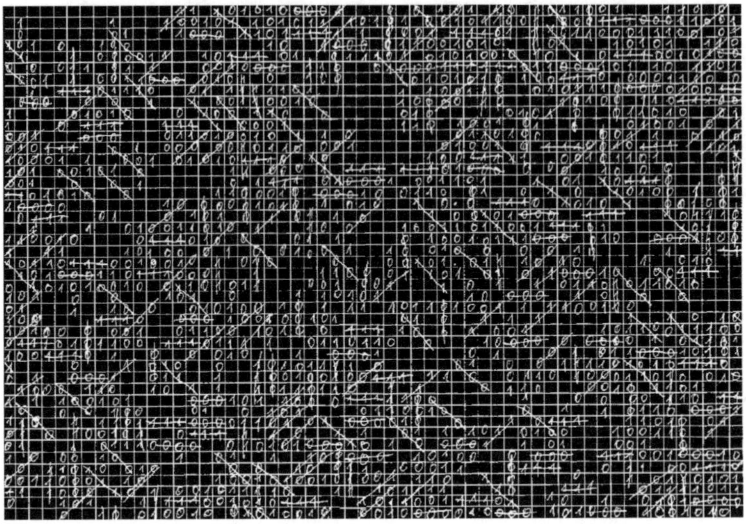

FIGURE 9.4 W. Gołuch, *Infinite drawing*, 2011

themselves, and our experience-based intuitions do not withstand confrontation with strict logical reasoning. Aporias, similar to those pointed out by Sextus Empiricus in relation to the concept of a point structure of a straight line, are also revealed when analyzing other issues discussed in this article:

1. We know that dichotomous divisions of a section will never lead us to a dimensionless point. But, at the same time, if we take seriously the geometric principle of infinite divisibility of a line section, we cannot assume that, in this way, we could obtain the smallest indivisible parts of a section. This is because the possibility of reaching such parts would contradict the basic assumption of the accepted principle. Thus, neither a dimensionless point nor a spatial "atom" can be the effect of dichotomous divisions.
2. The method of indicating points by cutting a straight line, accepted by mathematicians, is incompatible with our intuitive understanding of the continuity of a straight line, which we identify rather with a certain cohesiveness and in no way with fragmentation. Although thanks to the cuts, we could theoretically indicate an infinite number of points on a straight line and consequently reconstruct the entire straight line from them, it seems absurd to say that a continuous straight line would consist only of cuts (Łukowski 2011, p. 179).
3. The definition of an infinite set was based on the principle of similarity, which is rooted in direct experience. However, the very concept of actual infinity eludes not only our intuitions but also scientific observations of nature.

Therefore, to avoid falling into all sorts of contradictions, it would be most reasonable to agree that we cannot assume complete correspondence between the abstract constructions of mathematicians and direct experience, which includes both our perceptions and imaginations. Such a thought was expressed by Henri Poincaré, who stated that "the real mathematical continuum is quite different from that of the physicists and from that of the metaphysicians" (1905, p. 18).

Finally, it is worth considering the language and concepts we use to describe reality. When analyzing phenomena, we used to make sharp distinctions into what is continuous or discrete, single or plural, constant or variable, simple or complex. However, such extreme oppositions do not necessarily correspond to the objective properties of described things. This is because they may prove a simplified and selective way of perceiving reality, resulting from the deficiencies of our cognitive apparatus. Thus, they would point not so much to the features of the world itself but to the properties and limitations of its non-objective observers, which, after all, we are.

Notes

1 A discussion of various proposals for solving this paradox can be found, for example, in: (Łukowski, 2011).

2 Although the concept of the point structure of a straight line was already considered in antiquity, it was not widely accepted in mathematics and was also not part of the dominant trend in philosophy. A strict mathematical description of the point structure of a line (curve) was to appear only in *Geometry* by Descartes (Błaszczyk & Mrówka, 2014, p. v).

3 It does not mean, of course, that we know precisely how many of these elements there are since there are infinitely many of them. Rather, it is about the same *cardinality* – a measure of the number of the set's elements.

4 $25^{1\,312\,000}$ (Jernajczyk, 2018, p. 215).

References

Arnheim, R. (1969). *Visual thinking*. Berkeley.

Aristotle. (1930). *Phisica* (R. P. Hardie & R. K. Gaye, Trans.). Oxford.

Aristotle. (2016). *Metaphysics* (C. D. C. Reeve, Trans.).

Błaszczyk, P., & Mrówka, K. (2014). *Metafizyka ruchu w Geometrii Kartezjusza* (Argument Vol. 4).

Bolzano, B. (1950). *Paradoxes of the infinite* (Fr. Prihonsky, Trans.).

Borges, J. L. (1964). *The library of Babel* (J. E. Irby, Trans. [in:] idem, *Labyrinths*, pp. 58–64).

Cantor, G. (1877, June 25). *The letter from Cantor to Dedekind* (*Halle*), [in:] W. B. Ewald, *From Kant to Hilbert Volume 2: A Source Book in the Foundations of Mathematics*, New York 1996, pp. 856–860.

Davis, J. P., & Hersh, R. (1981). *The mathematical experience.*

Dedekind, R. (1972). *Continuity and irrational numbers* (W. W. Beman, Trans., [in:] idem, *Essays on the Theory of Numbers*, Chicago 1901, pp. 1–13).

Dedekind, R. (1888). *The nature and meaning of numbers* (W. W. Beman, Trans., [in:] idem, *Essays on the Theory of Numbers*, Chicago 1901, pp. 14–58).

Descartes, R. (1985). *Optics* (R. Stoothoff, Trans., [in:] idem, *The Philosophical Writings*, Bath).

Euclid. (1847). *The first six books of the elements of euclid* (O. Byrne, Trans.).

Galileo Galilei. (1914). *Dialogues concerning two new sciences* (H. Cew & A. De Salvio, Trans.) (Original work published 1638)

Gruszczyński, R. (2016). *Niestandardowe teorie przestrzeni* [*Non-standard theories of space*].

Jernajczyk, J. (2013). *Lossy representations of excessive reality*. In Ł. Huculak, et al. (Ed.), *Excess & lack* (pp. 172–176).

Jernajczyk, J. (2015). Irrational images – the visualization of abstract mathematical terms. *Mathematica Applicanda, 43*(2). 269–279.

Jernajczyk, J. (2016). Movement out of stillness – deliberations on the mechanism of creating a moving picture. In W. Gołuch & J. Jernajczyk (Eds.), *A picture set in motion* (pp. 82–99).

Jernajczyk, J. (2018). Ziarna myśli – o własnościach dyskretnych form reprezentacji informacji [Grains of thought – on the properties of discrete forms of information representation]. InR. Murawski & J. Woleński (Eds.), *Problemy filozofii matematyki i informatyki* [*Problems of philosophy of mathematics and computer science*] (pp. 213–222).

Kirk, G. S., & Raven, J. E. (1975). *The presocratic philosophers.*

Lucretius. (2001). *On the nature of things* (M. F. Smith, Trans.).

Łukowski, P. (2011). *Paradoxes.*

Nicholas of Cusa. (1996). *The layman on mind* (J. Hopkins, Trans.).

Nicholas of Cusa. (2000). *The bowling-game* (J. Hopkins, Trans.).

Poincaré, H. (1905). *Science and hypothesis.*

Proclus. (1992). *A commentary on the first book of Euclid's Elements* (G. R. Morrow, Trans.).

Russell, B. (1917). Mathematics and the metaphysicians. In *Mysticism and logic.*

Russell, B. (2009). *Our knowledge of the external world.*

Sextus Empiricus. (2018). *Against those in the disciplines* (R. Bett, Trans.).

Smolin, L. (2001). *Three roads to quantum gravity.*

CHAPTER 10

Simplifying Complexity

Copper(I) Complexes with Phosphines Derived from Peptides to Enhance Cancer Treatment

Urszula K. Komarnicka and Ewa Błaszczak

Abstract

In the words of Paracelsus, "All things are poison, and nothing is without poison; but the dose makes it clear that a thing is not a poison".[1] This concept resonates deeply in the field of cancer treatment, where new chemicals hold promise for killing cancer cells. However, the complexity arises from their lack of selectivity, affecting healthy cells alongside cancerous ones. This non-specificity leads to numerous side effects that can severely diminish the patient's quality of life. To address this issue, there is an urgent need for simple systems that can selectively target cancer cells. An approach that involves selective transport studies of chemotherapeutic agents is based on specific tumor cells' properties, such as the overexpression of various receptors. Cancer cells can utilize these receptors to bind drug carriers like carbohydrates, proteins, or peptides. By linking peptides known to bind these receptors with, for instance, copper complexes, researchers aim to deliver therapeutic compounds selectively to tumors, inducing their demise. Recent studies have synthesized phosphine ligands along with copper(I) complexes. These compounds were evaluated for their cytotoxic effects on various cancer cell lines. In the studied cancer cells, they accumulated predominantly in mitochondria and nuclei. This accumulation led to the generation of reactive oxygen species, a critical pathway contributing to their cytotoxic effects. Since cancer is a heterogeneous disease, the treatment landscape for cancer is complex and with challenges such as non-specificity, side effects, and drug resistance, the development of targeted delivery systems using peptide-linked copper complexes, or other organometallic complexes represents a promising step towards more effective and selective cancer therapies.

Keywords

human health – cancer – therapy – copper – peptides – inorganic chemistry

1 Introduction

Mortality from cancer may surpass that from cardiovascular diseases shortly. Approximately seven million people die from cancer each year, and it is estimated that there will be more than 18 million new cancer cases annually by 2025 (Siegel et al., 2021). Cancer, being a complex, heterogeneous disease is characterized by uncontrolled cell division. After primary tumor formation, cancer cells can further invade other tissues in the process termed metastasis (the spread of cancer to other body parts) (Vogelstein & Kinzler, 2004). To date, chemotherapy remains one of the major approaches to treating cancer by delivering cytotoxic substances to the cells. However, conventional chemotherapy has many disadvantages, such as the inability to deliver the correct amount of drug directly to cancer cells without affecting 'normal' cells. Drug resistance, altered biodistribution, biotransformation, and drug clearance are also common problems (Kakde et al., 2011). Hence, targeted chemotherapy and drug delivery systems are emerging as powerful methods to address these challenges. Such an approach enables selective and efficient delivery of drugs to specific targets, such as overexpressed receptors on cancer cells, while minimizing exposure to 'normal' cells (Aina et al., 2002; Dorsam & Gutkind, 2007; Howl et al., 2007; Meng et al., 2012; Zhang et al., 2012). For instance, tumor cells often express growth receptors like VEGFR (vascular endothelial growth factor receptor) and EGFR (epidermal growth factor receptor) exclusively on their surface (Padró et al., 2002). Various carriers such as carbohydrates, proteins, or peptides can selectively bind to these receptors (Allen, 2002; Duncan et al., 2005; Haag & Kratz, 2006; Hoskin & Ramamoorthy, 2008; Hynes, 1992; Janin, 2003; Nazarenko et al., 2013; Vicent, 2007).

2 Peptide-Mediated Delivery of Therapeutic Agents

Peptide-mediated delivery of therapeutic agents can be categorized into three distinct classes: (a) homing peptides (HPs), which have no internalization properties and therefore only deliver their cargo to the cell surface; (b) peptides that are linked to a cell-penetrating peptide (CPP), which enables cargo internalization via endocytosis or pore formation; and (c) cell-penetrating homing peptides (CPHPs), which can internalize without the aid of external agents. Several approaches for generating HPs with affinity for tissue-specific markers have been reported (Laakkonen & Vuorinen, 2010; Sugahara et al., 2010). As our understanding of peptide functionality and therapeutic effects continues to advance, peptide-based treatments are expected to play a crucial role in future therapeutic strategies. With ongoing research into the biological applications of polymeric materials, methods that involve peptide conjugation or physical encapsulation are likely to gain prominence in peptide delivery for both disease prevention and treatment (Du & Stenzel, 2014).

Numerous examples from the literature highlight peptide conjugates with organic therapeutics. Well-known anticancer drugs such as daunorubicin, doxorubicin, methotrexate, fluorouracil, and paclitaxel are frequently linked to carriers such as peptides or proteins (Fuertes et al., 2003; Haag & Kratz, 2006; Hudecz et al., 2005). For instance, two peptides that have entered clinical trials from the third CPHP group are the tumor-homing peptide RGD (Arg-Gly-Asp; the first CPHP that has undergone Phase I and II, and Phase III trials have been initiated), and NGR peptide (Asn-Gly-Arg). The latter has been tested in Phase I and II trials, targeting human tumor necrosis factor (hTNF) and enhancing doxorubicin delivery for refractory or resistant solid tumors (Eskens et al., 2003; Lorusso et al., 2012; Zhou et al., 2022). RGD and NGR peptide motifs selectively recognize integrins, i.e., proteins responsible for the growth, division, adhesion, and migration of cancer cells. These systems typically exhibit higher efficacy against tumor cells compared to 'normal' cells (Buckley et al., 1999; Majumdar & Siahaan, 2012; Sioud & Mobergslien, 2012; Zhu et al., 2020).

3 The Complexity of Inorganic Compounds

The compounds with transition metal ions offer vast opportunities for designing new therapeutics, beyond the scope of organic chemistry. Metal ions exhibit a wide range of geometries, oxidation states, and coordination numbers, allowing for precise tuning of their reactivity. Many transition metal compounds are being explored as alternatives to platinum-based chemotherapeutic agents. An interesting example of such complexes includes phosphine-diimine copper(I) complexes, known for their robustness and diverse biological properties such as anticancer, antibacterial, antiviral, antifungal, and anti-inflammatory activities, making them promising candidates for therapeutic applications (Bhattacharjee et al., 2021; Bykowska et al., 2014, 2018; Guz-Regner et al., 2020; Komarnicka et al., 2016, 2020; Kyzioł et al., 2018). Furthermore, the strong bond between the phosphine ligand and copper(I) prevents its oxidation to copper(II), a finding supported by previous studies (Bhattacharjee et al., 2021; Guz-Regner et al., 2020; Manna et al., 2019). Additionally, phosphine ligands can be easily functionalized. Notably, aminomethylphosphines derived from

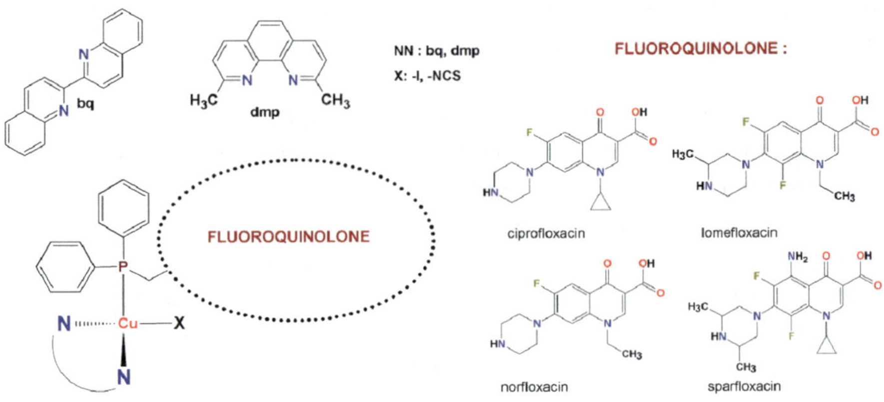

FIGURE 10.1 Copper(I) complexes with phosphines derived from fluoroquinolones

amino acids or prepared from highly water-soluble aliphatic secondary amines are promising for forming conjugates with a wide range of biomolecules (Komarnicka et al., 2015, 2018, 2021).

In Dr. Komarnicka's group, a series of stable iodide or thiocyanate copper(I) complexes with phosphine ligands derived from fluoroquinolone antibiotics such as ciprofloxacin (HCp), norfloxacin (HNr), lomefloxacin (HLm), and sparfloxacin (HSf) (Figure 10.1) have been synthesized and their physicochemical properties and biological activity have been characterized (Komarnicka et al., 2018, 2020, 2021).

To date, it has been shown that copper(I) complexes with phosphine-fluoroquinolone conjugates exhibit much higher cytotoxic activity than unmodified fluoroquinolone antibiotics, their phosphine derivatives, and even the well-known anticancer drug cisplatin (Figure 10.1). These same copper(I) inorganic compounds can induce apoptosis predominantly, regardless of cell type and incubation time. Additionally, electrochemical studies revealed that copper complexes with phosphine ligands derived from fluoroquinolone antibiotics could generate reactive oxygen species (ROS), as confirmed in cell culture experiments. The copper(I) complexes' anticancer activity is attributed to their interactions with DNA, causing breaks in the sugar-phosphate backbone and redox reactivity associated with ROS generation. Unfortunately, these compounds were toxic to 'normal' cells (Bykowska et al., 2018; Komarnicka et al., 2016, 2018, 2020, 2021). However, after encapsulation in liposomes, the water solubility and therapeutic index of the copper(I) complexes significantly increased

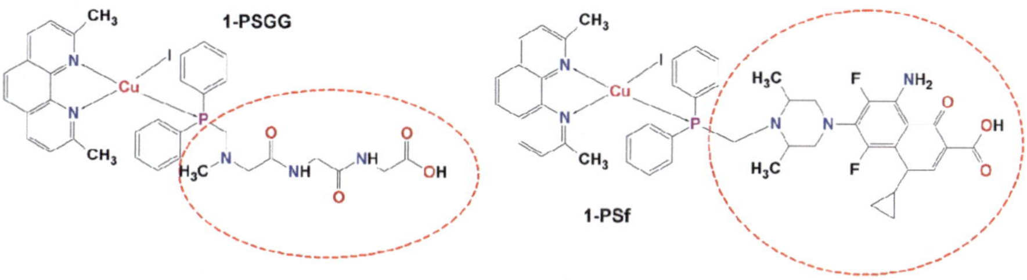

FIGURE 10.2 Copper(I) complexes with different phosphine ligands (differences highlighted with a circle; description in text)

(Kyzioł et al., 2018). It has also been proven that the cytotoxic effect of the described Cu(I) coordination compounds with derivatives of fluoroquinolones is dependent on the type of antibiotic. The absence of a methyl group on the piperazine ring and the presence of fluorine atoms in fluoroquinolone structures may contribute to the increased activity. Unfortunately, the high toxicity of these complexes has been a significant challenge (Bykowska et al., 2014; Komarnicka et al., 2020).

To reduce high toxicity towards 'normal' cell lines, antibiotic molecules were replaced with the water-soluble, simple, small, and inexpensive dipeptide motif Sar-Gly (HSG; Figure 10.2). It is worth mentioning that ^{11}C-glycylsarcosine (^{11}C-Gly-Sar) has been reported as a PET tracer, which can target H+/peptide transporters (PEPTs) functionally expressed in some human cancer cell lines. Additionally, these peptides offer well-known advantages as drugs, including specificity, potency, and low toxicity (Komarnicka et al., 2018, 2020, 2021).

Dr. Komarnicka's group also synthesized the diphenylphosphinomethyl derivative ($P(CH_2SG)Ph_2$; PSG), its oxide ($OP(CH_2SG)Ph_2$; OPSG), and the corresponding copper(I) complex (1-PSG) with the phosphine-peptide conjugate (PSG) (Figure 10.3). Additionally, they obtained a new starting salt (p-OCH_3-Ph)$_2$P(CH_2OH)$_2$Cl(MPOHC), the derived phosphine ligands P(p-OCH_3-Ph)$_2$-CH_2-OH (MPOH) and P(p-OCH_3-Ph)$_2$-CH_2-SarGly-OH (MPSG), and the corresponding copper(I) complexes [Cu(I)(dmp)MPOH] (1-MPOH) and [Cu(I)(dmp)PSG] (1-MPSG) (Figure 10.3). The intention was to increase the cytotoxicity of the copper(I) complexes with the phosphine-peptide conjugate by introducing an –OCH_3 group into the phosphine phenyl rings of P(p-OCH_3-Ph)$_2$-CH_2-SarGly-OH (MPSG). It is well known that substitution at the 4-position of the phenyl ring can significantly enhance the cytotoxicity of several therapeutics (Komarnicka et al., 2018, 2020, 2021).

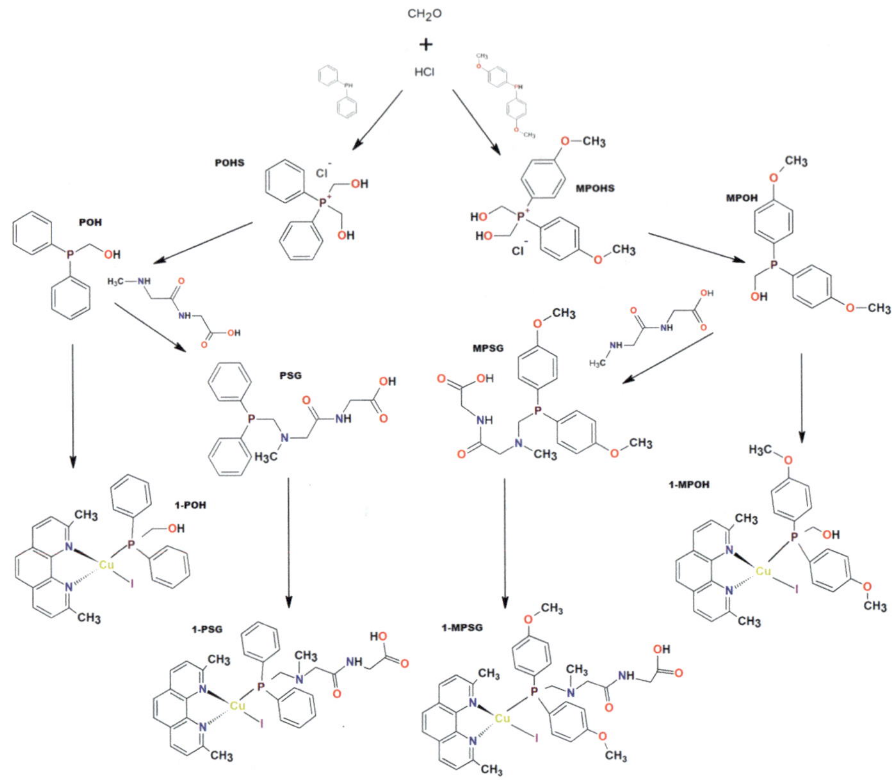

FIGURE 10.3 Schematic route of phosphine ligands with and without peptide, as well as copper(I) complex with phosphine ligands

4 The Biological Assessment of Synthesized Complexes

The cytotoxic activity of all presented organic and inorganic compounds, as well as cisplatin, was tested against several cancer cell lines: mouse colon carcinoma (CT26), human lung adenocarcinoma (A549), human breast adenocarcinoma (MCF7), human pancreatic cancer (PANC-1), and human prostate cancer (DU-145). Additionally, it was tested against several 'normal' cell lines: primary human pulmonary fibroblasts (MRC-5), human embryonic kidney (HEK293T), and human keratinocyte (HaCat). The obtained results indicate that the presence of $-OCH_3$ groups in the phenyl rings of the phosphine significantly increased the cytotoxic activity of the 1-MPOH and 1-MPSG complexes compared to 1-POH and 1-PSG complexes against the selected human cancer cell lines (MCF7, PANC-1, and DU-145) (Komarnicka et al., 2018, 2020, 2021).

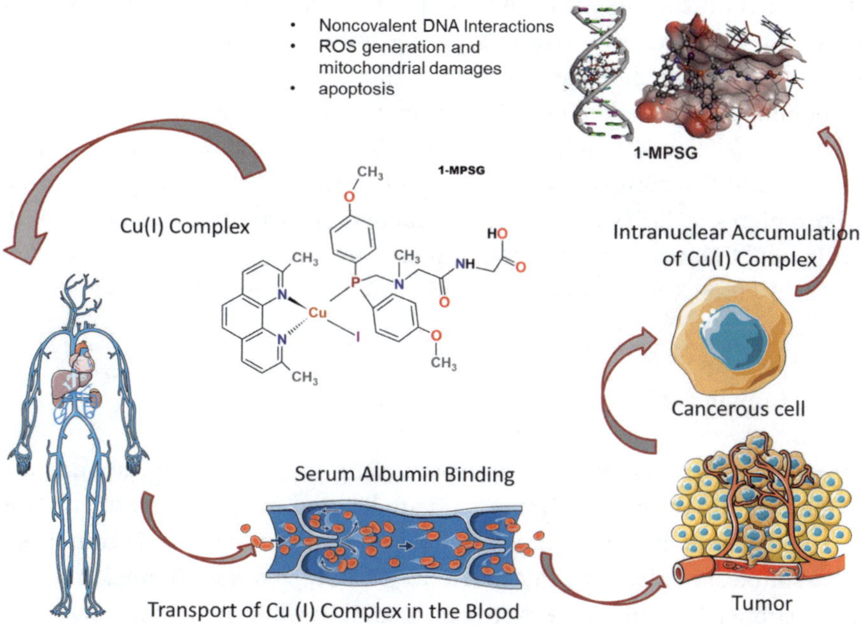

FIGURE 10.4 Schematic mechanism of cytotoxic action of copper(I) complexes with phosphine-peptide conjugate

A detailed study elucidated that complex 1-PSG accumulates significantly within MCF7 cancer cells, enhancing its cytotoxicity. Importantly, it was demonstrated that intracellular accumulation of 1-PSG was much higher in cancer cells compared to 'normal' cells. The attachment of the dipeptide Sar-Gly to the cytotoxic copper(I) complex via the phosphine motif increased the selectivity of this inorganic compound. Furthermore, 1-PSG induced apoptotic cell death in MCF7 cells, accompanied by a decrease in mitochondrial membrane potential and an increase in caspase-9 and caspase-3 activity. The complexes' ability to generate reactive oxygen species (ROS) in cells was also examined, confirming oxidative DNA damage (Figure 10.4) (Komarnicka et al., 2018, 2020, 2021).

We have begun exploring the antitumor potential of a new class of copper(I) complexes containing a diimine (dmp) and phosphine ligands $P(p\text{-}OCH_3\text{-}Ph)_2CH_2OH$ and $P(p\text{-}OCH_3\text{-}Ph)_2CH_2SarGly$, which feature an $-OCH_3$ group in the phenyl rings and the Sar-Gly peptide motif. The steric and electronic properties of the $-OCH_3$ group significantly influence the cytotoxic activity of these copper(I) complexes. Interestingly, the copper complex 1-MPSG exhibited increased cytotoxicity against MCF7, PANC-1, and DU-145 cell lines, while showing reduced cytotoxicity against CT26 and 7649 tumor cell lines. This suggests activation of different mechanisms compared to the analogous complex 1-PSG lacking the $-OCH_3$ motif on the phenyl ring. Importantly, the presence of both the Sar-Gly peptide motif and methoxy groups in the structure of copper(I) compound 1-MPSG resulted in high specificity and selectivity for MCF7, PANC-1, and DU-145 cell lines. Therefore, 1-MPSG can be considered the lead compound among all studied complexes due to its potent cytotoxic action, characterized by both high specificity and selectivity. Its notable activity is associated with multimodal interactions with DNA, primarily involving groove binding (Bykowska et al., 2018; Komarnicka et al., 2016, 2018, 2020, 2021).

5 Conclusions

This chapter presents a simple concept of using copper(I) complexes and the complex nature of their derivation and application in cancer treatment. It is evident that substituting fluoroquinolone molecules in copper(I) complexes with a simple, lipophilic peptide molecule (Sar-Gly) markedly reduces the toxicity of copper(I) complexes towards 'normal' cells while enhancing their cytotoxicity against cancer cells. This design strategy, involving cytotoxic copper(I) complexes with peptides attached via a phosphine linker, holds considerable promise for anticancer therapy.

Cancer is a complex and heterogeneous disease, necessitating therapies that simplify patients' lives by reducing side effects. The development of targeted metal complexes, with increased specificity, represents a step towards achieving this goal, offering a more selective approach that minimizes harm to healthy tissues while effectively combating cancer cells. Notably, metal complexes have shown promise by inducing cancer cell death mechanisms, such as, for instance, ferroptosis (Dixon et al., 2012; Dixon & Pratt, 2023). Further exploration of these diverse mechanisms broadens the potential avenues for innovative treatments. What is more, future research in this direction holds promise for refining therapies, enhancing cancer treatment strategies, and improving patient outcomes.

Acknowledgements

U.K. acknowledges financial support from the National Science Centre Poland, grant no. 2016/23/D/ST5/00269 and the Ministry of Science and Higher Education (grants no. 1233/M/WCH/13 and 1500/M/WCH/15). U.K. and E.B. also acknowledge the financial support of the National Science Centre Poland, grant no. 2023/51/B/ST4/00355.

Note

1 https://en.wikiquote.org/wiki/Paracelsus

References

Aina, O. H., Sroka, T. C., Chen, M.-L., & Lam, K. S. (2002). Therapeutic cancer targeting peptides. *Biopolymers, 66*(3), 184–199.

Allen, T. M. (2002). Ligand-targeted therapeutics in anticancer therapy. *Nature Reviews. Cancer, 2*(10), 750–763.

Bhattacharjee, A., Das, S., Das, B., & Roy, P. (2021). Intercalative DNA binding, protein binding, antibacterial activities and cytotoxicity studies of a mononuclear copper(II) complex. *Inorganica Chimica Acta, 514*, 119961.

Buckley, C. D., Pilling, D., Henriquez, N. V., Parsonage, G., Threlfall, K., Scheel-Toellner, D., Simmons, D. L., Akbar, A. N., Lord, J. M., & Salmon, M. (1999). RGD peptides induce apoptosis by direct caspase-3 activation. *Nature, 397*(6719), 534–539.

Bykowska, A., Komarnicka, U. K., Jeżowska-Bojczuk, M., & Kyzioł, A. (2018). CuI and CuII complexes with phosphine derivatives of fluoroquinolone antibiotics – A comparative study on the cytotoxic mode of action. *Journal of Inorganic Biochemistry, 181*, 1–10.

Bykowska, A., Starosta, R., Komarnicka, U. K., Ciunik, Z., Kyzioł, A., Guz-Regner, K., Bugla-Płoskońska, G., & Jeżowska-Bojczuk, M. (2014). Phosphine derivatives of ciprofloxacin and norfloxacin, a new class of potential therapeutic agents. *New Journal of Chemistry = Nouveau Journal de Chimie, 38*(3), 1062–1071.

Dixon, S. J., Lemberg, K. M., Lamprecht, M. R., Skouta, R., Zaitsev, E. M., Gleason, C. E., Patel, D. N., Bauer, A. J., Cantley, A. M., Yang, W. S., Morrison, B., 3rd, & Stockwell, B. R. (2012). Ferroptosis: an iron-dependent form of nonapoptotic cell death. *Cell, 149*(5), 1060–1072.

Dixon, S. J., & Pratt, D. A. (2023). Ferroptosis: A flexible constellation of related biochemical mechanisms. *Molecular Cell, 83*(7), 1030–1042.

Dorsam, R. T., & Gutkind, J. S. (2007). G-protein-coupled receptors and cancer. *Nature Reviews. Cancer, 7*(2), 79–94.

Du, A. W., & Stenzel, M. H. (2014). Drug carriers for the delivery of therapeutic peptides. *Biomacromolecules, 15*(4), 1097–1114.

Duncan, R., Vicent, M. J., Greco, F., & Nicholson, R. I. (2005). Polymer-drug conjugates: Towards a novel approach for the treatment of endrocine-related cancer. *Endocrine-Related Cancer, 12*(Suppl 1), S189–S199.

Eskens, F. A. L. M., Dumez, H., Hoekstra, R., Perschl, A., Brindley, C., Böttcher, S., Wynendaele, W., Drevs, J., Verweij, J., & van Oosterom, A. T. (2003). Phase I and pharmacokinetic study of continuous twice weekly intravenous administration of Cilengitide (EMD 121974), a novel inhibitor of the integrins alphavbeta3 and alphavbeta5 in patients with advanced solid tumours. *European Journal of Cancer, 39*(7), 917–926.

Fuertes, M. A., Alonso, C., & Pérez, J. M. (2003). Biochemical modulation of Cisplatin mechanisms of action: enhancement of antitumor activity and circumvention of drug resistance. *Chemical Reviews, 103*(3), 645–662.

Guz-Regner, K., Komarnicka, U. K., Futoma-Kołoch, B., Wernecki, M., Cal, M., Kozieł, S., Ziółkowska, A., & Bugla-Płoskońska, G. (2020). Antibacterial activity and action mode of Cu(I) and Cu(II) complexes with phosphines derived from fluoroquinolone against clinical and multidrug-resistant bacterial strains. *Journal of Inorganic Biochemistry, 210*, 111124.

Haag, R., & Kratz, F. (2006). Polymer therapeutics: Concepts and applications. *Angewandte Chemie, 45*(8), 1198–1215.

Hoskin, D. W., & Ramamoorthy, A. (2008). Studies on anticancer activities of antimicrobial peptides. *Biochimica et Biophysica Acta, 1778*(2), 357–375.

Howl, J., Nicholl, I. D., & Jones, S. (2007). The many futures for cell-penetrating peptides: How soon is now? *Biochemical Society Transactions, 35*(Pt 4), 767–769.

Hudecz, F., Bánóczi, Z., & Csík, G. (2005). Medium-sized peptides as built in carriers for biologically active compounds. *Medicinal Research Reviews, 25*(6), 679–736.

Hynes, R. O. (1992). Integrins: Versatility, modulation, and signaling in cell adhesion. *Cell, 69*(1), 11–25.

Janin, Y. L. (2003). Peptides with anticancer use or potential. *Amino Acids, 25*(1), 1–40.

Kakde, D. D., Kakde, R., Patil, A., Jain, D., & Shrivastava, V. (2011). *Cancer therapeutics-opportunities, challenges and advances in drug delivery*. https://japsonline.com/abstract.php?article_id=252&sts=2

Komarnicka, U. K., Kozieł, S., Starosta, R., & Kyzioł, A. (2018). Selective Cu(I) complex with phosphine-peptide (SarGly) conjugate contra breast cancer: Synthesis, spectroscopic characterization and insight into cytotoxic action. *Journal of Inorganic Biochemistry, 186*, 162–175.

Komarnicka, U. K., Kozieł, S., Zabierowski, P., Kruszyński, R., Lesiów, M. K., Tisato, F., Porchia, M., & Kyzioł, A. (2020). Copper(I) complexes with phosphines P(p-OCH3-Ph)2CH2OH and P(p-OCH3-Ph)2CH2SarGly. Synthesis, multimodal DNA interactions, and prooxidative and in vitro antiproliferative activity. *Journal of Inorganic Biochemistry, 203*, 110926.

Komarnicka, U. K., Pucelik, B., Wojtala, D., Lesiów, M. K., Stochel, G., & Kyzioł, A. (2021). Evaluation of anticancer activity in vitro of a stable copper(I) complex with phosphine-peptide conjugate. *Scientific Reports, 11*(1), 23943.

Komarnicka, U. K., Starosta, R., Guz-Regner, K., Bugla-Płoskońska, G., Kyzioł, A., & Jeżowska-Bojczuk, M. (2015). Phosphine derivatives of sparfloxacin – Synthesis, structures and in vitro activity. *Journal of Molecular Structure, 1096*, 55–63.

Komarnicka, U. K., Starosta, R., Kyzioł, A., Płotek, M., Puchalska, M., & Jeżowska-Bojczuk, M. (2016). New copper(I) complexes bearing lomefloxacin motif: Spectroscopic properties, in vitro cytotoxicity and interactions with DNA and human serum albumin. *Journal of Inorganic Biochemistry, 165*, 25–35.

Komarnicka, U. K., Starosta, R., Płotek, M., de Almeida, R. F. M., Jeżowska-Bojczuk, M., & Kyzioł, A. (2016). Copper(I) complexes with phosphine derived from sparfloxacin. Part II: A first insight into the cytotoxic action mode. *Dalton Transactions, 45*(12), 5052–5063.

Kyzioł, A., Cierniak, A., Gubernator, J., Markowski, A., Jeżowska-Bojczuk, M., & Komarnicka, U. K. (2018). Copper(i) complexes with phosphine derived from sparfloxacin. Part III: Multifaceted cell death and preliminary study of liposomal formulation of selected copper(I) complexes. *Dalton Transactions, 47*(6), 1981–1992.

Laakkonen, P., & Vuorinen, K. (2010). Homing peptides as targeted delivery vehicles. *Integrative Biology: Quantitative Biosciences from Nano to Macro, 2*(7–8), 326–337.

Lorusso, D., Scambia, G., Amadio, G., di Legge, A., Pietragalla, A., De Vincenzo, R., Masciullo, V., Di Stefano, M., Mangili, G., Citterio, G., Mantori, M., Lambiase, A., & Bordignon, C. (2012). Phase II study of NGR-hTNF in combination with doxorubicin in relapsed ovarian cancer patients. *British Journal of Cancer, 107*(1), 37–42.

Majumdar, S., & Siahaan, T. J. (2012). Peptide-mediated targeted drug delivery. *Medicinal Research Reviews, 32*(3), 637–658.

Manna, S. C., Mistri, S., Patra, A., Mahish, M. K., Saren, D., Manne, R. K., Santra, M. K., Zangrando, E., & Puschmann, H. (2019). Synthesis, structure, DNA/protein binding, molecular docking and in vitro anticancer activity of two Schiff base coordinated copper(II) complexes. *Polyhedron, 171*, 77–85.

Meng, L., Yang, L., Zhao, X., Zhang, L., Zhu, H., Liu, C., & Tan, W. (2012). Targeted delivery of chemotherapy agents using a liver cancer-specific aptamer. *PLoS One, 7*(4), e33434.

Nazarenko, I., Rupp, A.-K., & Altevogt, P. (2013). Exosomes as a potential tool for a specific delivery of functional molecules. *Methods in Molecular Biology, 1049*, 495–511.

Padró, T., Bieker, R., Ruiz, S., Steins, M., Retzlaff, S., Bürger, H., Büchner, T., Kessler, T., Herrera, F., Kienast, J., Müller-Tidow, C., Serve, H., Berdel, W. E., & Mesters, R. M. (2002). Overexpression of Vascular Endothelial Growth Factor (VEGF) and its cellular receptor KDR (VEGFR-2) in the bone marrow of patients with acute myeloid leukemia. *Leukemia, 16*(7), 1302–1310.

Siegel, R. L., Miller, K. D., Fuchs, H. E., & Jemal, A. (2021). Cancer Statistics, 2021. *CA: A Cancer Journal for Clinicians*, *71*(1), 7–33.

Sioud, M., & Mobergslien, A. (2012). Selective killing of cancer cells by peptide-targeted delivery of an anti-microbial peptide. *Biochemical Pharmacology*, *84*(9), 1123–1132.

Sugahara, K. N., Teesalu, T., Karmali, P. P., Kotamraju, V. R., Agemy, L., Greenwald, D. R., & Ruoslahti, E. (2010). Coadministration of a tumor-penetrating peptide enhances the efficacy of cancer drugs. *Science*, *328*(5981), 1031–1035.

Vicent, M. J. (2007). Polymer-drug conjugates as modulators of cellular apoptosis. *The AAPS Journal*, *9*(2), E200–E207.

Vogelstein, B., & Kinzler, K. W. (2004). Cancer genes and the pathways they control. *Nature Medicine*, *10*(8), 789–799.

Zhang, X.-X., Eden, H. S., & Chen, X. (2012). Peptides in cancer nanomedicine: drug carriers, targeting ligands and protease substrates. *Journal of Controlled Release: Official Journal of the Controlled Release Society*, *159*(1), 2–13.

Zhou, M., Zou, X., Cheng, K., Zhong, S., Su, Y., Wu, T., Tao, Y., Cong, L., Yan, B., & Jiang, Y. (2022). The role of cell-penetrating peptides in potential anti-cancer therapy. *Clinical and Translational Medicine*, *12*(5), e822.

Zhu, L., Ding, Z., Li, X., Wei, H., & Chen, Y. (2020). Research progress of Radiolabeled Asn-Gly-Arg (NGR) Peptides for imaging and therapy. *Molecular Imaging*, *19*, 1536012120934957.

CHAPTER 11

The Complexity of the New Media

Katarzyna Kopecka-Piech

Abstract

This paper presents three levels of complexity with regard to the new media: methodological complexity, analytical complexity, and complexity of communication practice. The first of these refers to the diversity and hybridisation of research approaches. The second encompasses direct complexity (the application of complex systems theory) and indirect complexity (breaking the analysis down into small elements that fit various systems). The third, complexity of communication practice, is analysed using an example that illustrates its productive application. One of the research outcomes pertaining to this is the impact that the complexity of the new media has had on the nature of media studies.

Keywords

new media – media and communication studies – complexity – research methodology

1 New Media Are Complex

Today, new media are considered digital communication technologies. In the past, however, new media was understood to be any channel or platform of communication that introduced a qualitatively significant change. Since the spread of the Internet, new media has also been identified with the Internet itself. Regardless of the definition, the spread of the new media implies significant transformations at all stages and levels of communication.

The advances in the area of new media have brought about a time of multilevel convergence, hybridisation of the communication practice, and consequently more intricate research methods. New media are complex. At the macro level, they have a network structure, at the meso level they are characterised by structural complexity and at the micro level they are bound by

DOI:10.1163/9789004714540_011

the diversity of affordances (Hopkins, 2020) and content. This has many consequences. Firstly, monomedia practice is no longer commonplace although it still exists. The media now create a converging environment, in which the user is immersed. The users live their lives in media and not only with media (Deuze, 2012). This not only refers to the classic broadcaster or producer but even more so to the recipient or the consumer. Secondly, from the research perspective, the analysis escapes mono-disciplinary and mono-methodology approaches. The nature of the media and the consequences of its existence are too multi-faceted to be explained simply by looking at them from a single perspective and applying classic tools. New media call for multi-faceted research with regard to the disciplines, methodology, methods and theory as well as to day-to-day analytical activities. To research media is to live with the media and there are certain consequences to that.

This study aims to present three aspects of the new media complexity, which have been ordered by the level at which the media reality is perceived. It begins with the meta-level of perceiving the world, then moves to analytical structures and ends with communication practice. Firstly, this study points to methodological complexity, which on the one hand refers to the diversity of ways of looking at the same issue and the criss-crossing of various approaches, which brings about the emergence of a new discipline, or rather trans-discipline. Research into new media contributes much more greatly to the multidisciplinary character of media and communication studies than research into traditional media. Secondly, this study describes the complexity of research into new media resulting from their network and algorithmic structure, language and clear distinctiveness. Finally, the complexity of the new media communication practice is analysed. How is the era of linear monothematic and often single-authored communication through words, pictures and sound coming to an end? What consequences does it cause? This paper highlights the emergence of a new (pre-)paradigm.

2 Methodological Complexity

A handbook might be considered a type of publication, which as a compendium of knowledge, embraces the most relevant issues regarding its subject and so gives the reader an overview of a given subject. Even a perfunctory look through handbooks on new media makes one realise the complexity of the nature of those media and of their tendency to become even more complex (particularly through hybridity and convergence). This trend by representatives of multiple disciplines can be illustrated by the evolution of new media handbooks.

We can start with Andrew Dewdney and Peter Ride's 2014 *The Digital Media Handbook*, which is an update on 2006 *The New Media Handbook*. The change in the title itself is significant: it is hard to talk of new media when the old and new media have become converged. The handbook exclusively discusses digital media but it is worth noting that convergence also applies to analogue media, such as the press. The first version of the handbook had four chapters: *Frames of New Media, New Media Practice, Forms of New Media* (including such issues as the interface, interactivity and code) and *New Media Theory and Practice* (which only included four pages on convergence). The 2014 version has five chapters: *Networks, Convergent Media, Creative Industries, Digital Media, Media Histories and Theories.* In only a few years, the media reached a new stage of development which resulted in a complete shift in analytical perspective. It is this media complexity that now has the most significance – its network structure and convergence on various levels (whether economic, technological or cultural) and interdisciplinary phenomena, such as the expansion of creative industries. The media are no longer a synonym for a broadcaster (which was the case in classic media concepts, e.g. the role of television). The media are an environment in which producers and consumers, or prosumers (Toffler, 1997) and producing users (*produsers*) (Burns, 2008) operate.[1]

Let us now look at a handbook published a dozen years later, the 2021 *Routledge Handbook of Digital Media and Communication* edited by Leah Lievrouw and Brian Loader. The table of contents covers 27 chapters in a three-part framework: *Artifacts, Practices* and *Arrangements.* What we are confronted with in attempting this synthesis of key themes, topics and issues is an enormous dispersion and diversity of levels of understanding and analysis. From issues emphasising technology (specific infrastructures, mobile media, social media), through the main categories characterising contemporary new media (big data and algorithms, ubiquity, surveillance, hybridity, mediatization, globalization), to key new media practices (hacking, archiving, identity building) and their contexts (democracy, economics and finance, technology, war, government, the public sphere). Understanding new media nowadays requires an intersection of orders and a truly interdisciplinary reflection.

There are many indications that the next iterations of technological innovations, such as artificial intelligence, will further expand and fragment new media discourse. But at the same time, such a complex phenomenon as new media cannot be researched in a single-track fashion as this would result in an incomplete explanation of the situation. It is becoming increasingly common for representatives of multiple disciplines to research one issue (such as privacy and surveillance on the web, blurring distinctions between free time and work online, etc.) and confront the results, or for single researchers to carry out interdisciplinary projects. The key to media research complexity is the search for new theories and concepts from diverse disciplines, which would be applicable in the research into a given issue. Just as mathematic theories were used in the past to create communication models,[2] today informatics is used to explain the specificity of new media. New media cause media studies to branch out in many different directions. Their technological background as well as cultural, social, political and economic significance are related to notions from such fields of study as psychology, education or linguistics (to name only a few disciplines, which are significant for media studies). Formally, in countries such as Poland, it makes media studies a multidiscipline or an inter-discipline, which has certain administrative and legal implications (Jabłonowski & Gackowski, 2012).

The methodological complexity not only refers to the diversity of disciplines and research perspectives but also to the very theory of media and communication studies. Media studies are of hybrid-like as its history and tradition were born at the crossroads between science and the humanities. The methodology of new media research is an example of an interdisciplinary methodology. It applies IT terminology, humanist, social and economic theory, artistic visualisations as well as quantitative and qualitative methods and tools. There are many examples of interdisciplinary methodologies, i.e. Lev Manovich's approach (2006) incorporating elements of poetics, film studies and IT; or Henry Jenkins' (2006), which combines sociology, marketing and culture studies. We can therefore claim that due to the application of the complex systems theory, the new media methodology and theory are in themselves complex systems which, due to the nature of the research subject, are in a state of perpetual emergence, balancing between order and chaos.

3 Analytical Complexity

We consider analytical complexity as being indirect or direct. Indirect analytical complexity means the necessity to 'break down' a new medium to basic elements and factors for an in-depth analysis of its structure, functionalities, content and the impact coming from all those elements melting together. This also refers to the examination of processes, such as the production/creation and consumption/use of media, which are two basic, but not only, media processes. These processes are of a multi-stage character and depending on the assumed strategy and the model of value generation they are linear, non-continuous or network-like.[3] The multitude of concepts, terminologies and in particular typologies, makes it hard to reach a consensus on the terminology and analytical categories to be used. It is particularly the emergence of new media phenomena that results in new proposals being put forth.

The analytical complexity results directly from the complex nature of the media. On the one hand, a digital medium is a technology with a physical side (such as hardware and infrastructure) and a virtual side (such as the interface, software and content); on the other, it is a cultural artefact, but it may also be treated as an institution, organisation, consumer product, social environment, living space, etc. Depending on the notion being examined, different analytical categories will be applied but their specificity will remain similar. It is impossible to run an analysis without combining what used to be separate orders. Research of new media users analyses *produsage* (productive usage) (Bruns 2008), which is an analysis, blending elements of economics (production) and technology (usage); or work through play (*playbour*) (Kücklich, 2005). 'Clear-cut' analytical categories no longer reflect contemporary reality and fail to grasp the complexity of new media.

Analytical complexity may also be direct and mean the direct application of the complex systems theory, regarding for instance networks. Network theories perfectly match the specificity of contemporary media. The network structure is heterarchical (Kontopoulos, 1993). Unlike the hierarchy, the lower levels do not contain themselves within the higher levels but intersect with them. This is what networks, including communication networks, do. They connect the different levels giving flexibility to the whole (Van Dijk, 2006). Such is the network structure of e.g. content flow, the presence of media personalities or marketing relations between organisations. Additionally, all those networks create more complex networks of metarelations.

This makes network tools applied in media and communication studies justified and effective. Communication studies research stresses the significance of interpersonal networks for shaping opinions and relations between social networks and so-called mass media. What contributed to the growing interest in network phenomena has undoubtedly been the development of computer networks. Media and communication studies use the classic network analysis, which focussed on relations between people and market entities although it may also be used to analyse relations between IT texts (Schenk, 1995; after: Quandt, 2008). The application of network theories for Internet studies seems obvious. We have observed illustrating the structure ('the appearance') of the internet with graphs and sociological studies, including social network analysis, on relations in social media or strict media studies, e.g., research of media contents (Kopecka-Piech, 2011, 2013). During the last decade many IT-based methods and tools have been developed opening up an unlimited field of research not only into the structures of the network, nodes and relations but above all into its consequences. Nowadays it is difficult to imagine modern media and communication studies without the analysis of big data flow and its relationship with communicative actors, which include human and non-human subjects. The situation is becoming increasingly complicated putting posing the question and the limits of media studies.

While indirect analytical complexity is inevitable and its inclusion seems to be a necessity, direct complexity is a matter of choice and like stripping each methodology, causes certain effects. Limiting oneself to network analysis involves all kinds of constraints. Network analysis focuses on connections and not on what caused a particular configuration and hence needs to be supplemented with another perspective. This is where the issue of the necessary methodological complexity returns. A network is a special diagram providing topological information on the connected nodes, rather than information regarding them (Berry 2008). According to Beery (2008), networks favor synchronous perception at the expense of diachronic perception, level the importance of the past in favor of the present, and further accentuate what is in the network, i.e., what is connected, while somewhat ignoring what is outside the network, what is disconnected.

4 The Complexity of Communication Practice

New (and traditional) media form networks, not only in the form of the hypertextual construction of the Internet but also as network structures shaped by the flow of content and discursive, marketing and market practices (Kopecka-Piech, 2013). Network formation is an essential form of the cooperation strategy, image building carried out by media organisations and – in the era of hybridity and datafication (Meijas & Couldry, 2019) – creation and distribution of data and content. For professional creators of media content, this means converged practices while for the user 'reception' becomes a multi-platform, transmedia, and often multi-purpose and simultaneous experience (reading, listening and watching different media happens (virtually or actually) simultaneously, while creating one's content is a common activity, whether additional or primary) (Kopecka-Piech, 2011).

The aforementioned produsage is a good example to illustrate the complexity of communication practices, which requires the application of comprehensive analysis for its research. A good case in point are social media. They are means of communication between users as well as a marketing tool, which may be applied to such operations as building a brand. A distinctive way of categorising content on social media is the hashtag (#). Aggregating content makes it easier to reach the users interested in the subject and increases the number of recipients of the content created by the user.

What does productive usage look like for social media users? Let's take the example of Instagram to illustrate this. After downloading the application, the user can view the content created by their (added) friends or strangers disseminating their content, including owners of brands, products, services, etc. Productive usage already happens at this stage: by viewing content, commenting and clicking on the heart icon to show approval the user participates in the complex, and in a way 'massive' act of producing a platform that is a source of income for its owners (e.g., through advertising). Actions taken by 'ordinary' users are also beneficial to brand owners. This stage and this aspect call for research involving such fields of study as economics, management (particularly marketing), sociology, culture studies and sometimes also linguistics and psychology. To obtain a complete picture, it should encompass an analysis of its technical, textual and visual layer, as well as research into the social, cultural and economic significance of the users' activity.

Noticeably, a more advanced form of productive usage is the creation of content. This process is technologically complex because in addition to the basic ability to operate the smartphone or tablet (downloading and launching the application, using mobile internet, etc.) it requires the user to operate the pre-installed camera. To create a message, the user first needs to take a photo or make a video, then edit it using available filters, name it and tag the contents. Next, the user may add their location (using the GPS), pinning the photo to a place on the map and giving it a geographical aspect. Later, the user may tag their friends (to be recognised by other users) and publish the post on Instagram and possibly elsewhere. At this stage, the user's activity enters the more complex dimension of the interconnectedness of applications. The post may be also published on Facebook or X (former Twitter) and then, using the share feature, be included in the content 'created' by other users.

This exemplifies the network structure of online businesses, relations and technological links as well as the practice of stimulating consumption (or more precisely, prosumption) and building one's image through individual profiles. The content loses some of its single-authored characters and becomes more multi-authored, or even public. Research into this type of phenomenon is highly complex as it goes beyond the world of applications, the internet and other new media and into the user's living space, public and business sphere, etc. It is then that the effects and consequences of a particular structure and content are revealed. If a user starts to run their profile as a business, the content is no longer spontaneously created but becomes a methodical strategy. The trend of influencers forming opinions is growing, which has greatly redefined traditional sources of information and led to the emergence of the social media opinion leaders, so-called influencers. This shows how layers of media technologies, contents and activities overlap and form a variety of networks. The main outcome of this is the emergence of a new communication paradigm, or preparadigm, of multi-layered, multi-directional, non-linear, multi-authored and multi-purpose communication.

5 Complexity, Hybridisation, Dynamisation

The complexity of new media can therefore be considered widespread and present in many areas of their operation. It is accompanied by a multidimensional hybridisation of both the technology itself (between the online and offline worlds, materiality and virtuality) and various aspects of users' lives. It is increasingly based on balancing processes involving mediated human-to-human communication, human-to-machine communication, and communication with or based on artificial intelligence, increasingly becoming the communicative actor.

The activity is complex and complicated by nature, which does not necessarily entail difficulties in its implementation. Communication tools are becoming more and more intuitive and suited to people's needs and senses. Touch screens on smartphones, for example, have activated the sense of touch much more than hearing and sight. Moreover, new media are undergoing permanent dynamic transformation, perpetual evolution. This leads the way towards ever greater possibilities and means, e.g. at the technical level, growing complexity, but on the user's level more intuitiveness and adaptation to a wider range of users' daily activities. This offers great opportunities, but at the same time poses many ethical, ideological and political challenges. The nature of new media calls for a complex research methodology and tools from other disciplines and theories to fit the new areas of life being embraced by new omnipresent media, increasingly based on artificial intelligence. Consequently, media and communication studies will also evolve in new directions, resulting permanently in new, evolving (pre-)paradigms.

Notes

1 More on hybrid roles of media users in: K. Kopecka-Piech, 2014.

2 A classic example is Shanon and Weaver's model.

3 Asle Rolland (2009) indicates three strategies of value generation: chain, network and shop.

References

Berry, D. M. (2008). The poverty of networks. *Theory, Culture & Society, 7–8*(25), 364–372.

Bruns, A. (2008). *Blogs, Wikipedia, second life and beyond. From production to produsage*. New York, NY: Peter Lang.

Deuze, M. (2012). *Media life*. Cambridge: Polity Press.

Dewdney, A., & Ride P. (2006). *The new media handbook*. Abingdon, NY: Routledge.

Dewdney, A., & Ride, P. (2014). *The digital media handbook*. Abingdon NY: Routledge.

Hopkins, J. (2020). The concept of affordances in digital media. In H. Friese, M. Nolden, G. Rebane, & M. Schreiter (Eds.), *Handbuch soziale praktiken und digitale alltagswelten* (pp. 47–54). Wiesbaden: Springer.

Jabłonowski, M., & Gackowski, T. (2012). Tożsamość nauk o mediach. Obszary, perspektywy, postulaty. *Studia Medioznawcze*, 2(49), 15–24.

Jenkins, H. (2006). *Kultura konwergencji. Zderzenie starych i nowych mediów*. Warszawa: Wydawnictwo Akademickie i Profesjonalne.

Kontopoulos, K. M. (1993). *The logics of social structure*. New York, NY: Cambridge University Press.

Kopecka-Piech, K. (2011). *Media convergence strategies. Polish examples*. Wrocław: Astrum.

Kopecka-Piech, K. (2013). Usieciowienie mediów w świetle teorii grafów. Przykład stacji TVN. In K. Pokorna-Ignatowicz, S. Jędrzejewski, & J. Bierówka (Eds.), *Nowe media a praktyki komunikacyjne* (pp. 109–129). Kraków: AMF.

Kopecka-Piech, K. (2014). Prosumpcja, produkcyjne użycie (*produsage*), praca przez zabawę (*playbour*). Hybrydyczne zmiany relacji w Kulturze 2.0. In B. Walczak, A. Niekrewicz, & J. Żurawska-Chaszczewska (Eds.), *Komunikacja w stechnicyzowanym świecie. Wpływ postępu technologicznego na język i literaturę* (pp. 103–111). Gorzów Wielkopolski: Wydawnictwo Naukowe PWSZ im. Jakuba z Paradyża.

Kücklich, J. (2005). Precarious playbour: Modders and the gigital games industry. *The Fibreculture Journal*, (5). http://five.fibreculturejournal.org/fcj-025-precarious-playbour-modders-and-the-digital-games-industry/

Manovich, L. (2006). *Język nowych mediów*. Warszawa: Wydawnictwo Akademickie i Profesjonalne.

Mejias, U. A., & Couldry, N. (2019). Datafication. *Internet Policy Review, 8*(4), 1–10.

Quandt, T. (2008). Network analysis. In W. Donsbach (Ed.), *The international encyclopedia of communication*. Blackwell Reference Online. http://www.communicationencyclopedia.com/subscriber/tocnode?id=g9781405131995_chunk_g978140513199519_ss8-1

Rolland, A. (2009). Convergence as strategy for value creation. *International Journal on Media Management, 1*(5), 14–24.

Schenk, M. (1995). *Soziale Netzwerke und Massenmedien: Untersuchungen zum Einfluß der persönlichen Kommunikation*. Tübingen: Mohr.

Toffler, A. (1997). *Trzecia fala*. Warszawa: Państwowy Instytut Wydawniczy.

Van Dijk, J. (2006). *The network society: Social aspects of new media*. London: Sage Publications.

CHAPTER 12

The Simplicity and Complexity of Philosophical Discourse

Two Different Ways of Reasoning about the Stoic Postulate of Premeditation of Evils (*Praemeditatio malorum*)

Marek Krajewski

Abstract

Stoic philosophers (Epictetus, Seneca, Marcus Aurelius, and others) regarded the strategy of premeditatedly brooding on the evil that might befall us, "premeditation of evils" (*praemeditatio malorum*), as one of the ways of safeguarding ourselves against evil. Before misfortune strikes – they postulated – imagine it clearly with all its details, and then it will be less destructive. In this work, the author presents two methodologically different ways of justifying the postulate of *praemeditatio malorum*: a literary-persuasive and an analytical one. Both of them are characterized by their simplicity, but that very term, in each of these approaches, has a different meaning when taken on its own. In the first instance, simplicity means "ease of understanding a philosophical text", whereas in the second it designates "stylistic asceticism". Which simplicity should we choose? Each of the two approaches has its shortcomings: readily understandable reasoning is not always the same as logically coherent stylistic asceticism, while a dry, quasi-mathematical style limits the potential number of readers a text may have. The author, despite all his own objections, implicitly supports the second method.

Keywords

stoic philosophy – premeditation of evils – logical agrumentation

I will present and compare two contemporary ways of reasoning of the Stoic postulate of premeditation of evils. One of them by P. Stankiewicz (2014, pp. 219–235) is of literary-persuasive nature, and the other one (my own) – is an attempt at analytical reasoning. I will show the advantages and disadvantages of each argumentation separately in the context of the theme question of this volume that is complexity and simplicity. I will analyze the meaning of two nominal groups, namely *the complexity of the philosophical text* and *the simplicity of the philosophical text*. By "simplicity" in this article, I understand "the easiness of understanding a philosophical text", whereas by "complexity" I understand "the difficulty in understanding a philosophical text". I will answer the following question: which way of reasoning (literary-persuasive or analytical) is simpler/more complex and I will show the price of that simplicity/complexity.

1 The Premeditation of Evils in Stoic Philosophy

Ancient Stoic philosophers put ethics at the center of their interests. They understood it as theoretical science about "what happiness is" and as "the art of living". They believed that it indicated "by what means it [happiness M.K.] can be achieved" (Reale, 2010, p. 394).

In that second practical understanding ethics became a set of advice, didactical examples and spiritual exercises, presented to all those who wanted to live happily. One of those exercises was daily meditation which consisted of imagining various (also negative) life events. Amongst others, its aim was to prepare yourself for "the moment when unexpected circumstances occur, maybe dramatic ones" (Hadot, 2003, p. 20).

That was the premeditation of evils (*praemeditatio malorum*; hereinafter I will use the PM abbreviation) recommended by Stoics in order to avoid surprise and face difficult situations (Hadot, 2003, p. 148). Let that mental exercise be characterized now by one of the most renowned Roman Stoics Lucius Annaeus Seneca (Moral Letters to Lucilius 91, 3–4. 8; in: Seneca, 1961, pp. 435–437):

> for it is the unexpected that puts the heaviest load upon us. Strangeness adds to the weight of calamities, and every mortal feels the greater pain as a result of that which also brings surprise. Therefore, nothing ought to be unexpected by us. Our minds should be sent forward in advance to meet all problems, and we should consider, not what is wont to happen, but what can happen [...] Exile, the torture of disease, wars, shipwreck, – we must think on these [...] Let us place before our eyes in its entirety the nature of man's lot, and if we would not be overwhelmed, or even dazed, by those unwonted evils, as if they were novel, let us summon to our minds beforehand, not as great an evil as oftentimes happens, but the very greatest evil that possibly can happen. We must reflect upon fortune fully and completely. (http://www.stoics.com/seneca_epistles_book_2.html. Translated by Richard M. Gummere)

The result is the feeling of mental power and bravery (Seneka, On Anger 3, 37, 3; in: Seneka, 1989, p. 388):

> Make up your mind that there are many things which you must bear. Is any one surprised that he is cold in winter? That he is sick at sea? That he is jolted about on the highroad? The mind will meet bravely everything for which it has been prepared. (http://www.stoics.com/seneca_essays_book_1.html#ANGER1. Translated by John W. Basore)

Along with the renaissance of the stoicism of which we are witnesses (compare Becker, 1999; Irvine, 2009, 2013; Fabjański, 2010, 2014; Mazur, 2010, 2014; Stankiewicz, 2014), it is desirable, in my opinion, to provide a convincing substantiation of PM as an appropriate strategy of conduct in everyday life.

That burning need comes from the fact that PM in a natural way contradicts the view very popular since the 50s and still advocated on the effectiveness of the *power of positive thinking* – which is the title of a famous psychological self-help book by Pastor Norman Vincent Peale. Necessarily, argumentation in favor of PM has to take into consideration the main theses of the supporters of the competitive idea of positive thinking (which I call here 'imagined premeditation of fortunate events', *praemeditatio bonorum*, abbreviated to PB).

2 Literary-Persuasive Substantiation of PM

The book by P. Stankiewicz (2014) has features of a psychological-philosophical handbook and it is written in a style adequate for the aim which was formulated *expressis verbis* in its very beginning: "The book you are holding in your hands has a simple yet ambitious aim: to teach you how to live happily" (Stankiewicz, 2014, p. 5).

That clear declaration in the Introduction suggests that the reader will, later on, be educated and essentially encouraged to carry out appropriate conduct. And it actually is that way. The author consistently tends to convince the reader about the benefits which come from a stoical way of life and one of the means leading to that is subsequent persuasion given in a brilliant, literary form, full of various rhetorical tricks. That is why the substantiation of P. Stankiewicz I call "literary-persuasive".

The philosopher devotes an entire chapter to the issue of PM which he calls "premeditation of evils" (Stankiewicz, 2014, pp. 219–235). Due to the to the space limitations, I do not quote the whole text by P. Stankiewicz but I limit myself to presenting an argumentation outline (see more on that topic in Kisielewicz, 2014, pp. 172–191), presented in schematic narration. It is divided into theses (sometimes supported by additional presuppositions) which are followed by arguments supporting the theses. Stankiewicz also presents anti-theses and counter-arguments by which these anti-theses are overturned. The outline is as follows.

Thesis I: We have to foresee each evil or a worse variant of a possible situation (for example in a 'win-lose' situation I chose the second variant, amongst many variants I chose the worst one) because

Six arguments to support Thesis I:

1. we will not be surprised by evil,
2. we will 'disarm' evil in advance,
3. surprise may only be positive (a black scenario will either come true and we will avoid the surprise or it will not come true and we will be pleasantly surprised),
4. we will avoid the feeling of biting uncertainty and the pain of awaiting the finale,
5. PM is a passive defense against the predominating forces of fate and defense is more effective if we are ready for attack (Additional presumption: life is a battle, the favorite metaphor of ancient stoics), and
6. PM is an active defense against the predominating forces of fate, and the defense is more effective if we exercise in it often and we use it preventively (Additional presumption: life is a battle; see above).

Additional thesis I: You should imagine only painful and only probable evil because:

Two arguments supporting the additional thesis I:

1. it is impossible to imagine all possibilities,
2. an unlike (of small likelihood) event does not really worry us that much, and if we do not worry in advance that PM will not work *ex definitione*.

Additional thesis II: We imagine misfortune by visualizing them 'concretely and in detail' because only then PM will work.

Additional thesis II is not substantiated by any arguments.

Sub-thesis of the Additional thesis II: The effectiveness of PM is directly proportional to the efforts of imagination.

Sub-thesis of the Additional thesis II is not substantiated by any arguments.

Additional thesis III: PM also means creating emergency plans (I call it 'chess-player's strategy'). Chess-player's strategy is effective.

Additional thesis III is not substantiated by any arguments.

Anti-thesis I: PM is being a doomsayer and a pessimist, which infests our lives and is an excessive cost of PM.

Argument overthrowing anti-thesis I: being a doomsayer is different than PM in (1) the aim (a doomsayer is sad just to be sad and a stoic is sad for mental exercise), (2) control (a doomsayer on the contrary to a stoic does not control sadness), and (3) time scale of deliberations (a doomsayer unlike a stoic, does often, all the time, and one bad projection is driven by another).

Anti-thesis II: PM is 'speaking of the devil' and a bad projection can come true.

Argument overthrowing anti-thesis II: There's no cause-and-effect relationship between the projection and the actual, projected event happening.

Anti-thesis III: If I a misfortune will happen for sure, all I have left is waiting for death.

Argument overthrowing anti-thesis III: That belief is a caricature. PM does not cancel the rule of rational planning in accordance with the rules of probability. If the probability of a positive event is overwhelming, we plan our moves accordingly to that potential positive event, however, we have to remember all the time (although without a lot of worries) about its less probable (negative) alternative.

Anti-thesis IV: There are misfortunes of such slight probability that foreseeing them in accordance with Additional thesis I is a mistake. Nevertheless, when they do happen, we are right to blame PM and we decide that PM has some gaps.

Argument overthrowing anti-thesis IV: Despite everything, let us not forget about less probable misfortunes (for example tsunami). It is better than being an incurable optimist anyway.

Piotr Stankiewicz did not avoid two lapses.

First of all, he assumed that in the fifth and the sixth argument for the thesis I is an additional presumption (life is a battle) which is very disputable. (What justifies it apart from phrasal tradition?). The author supported it only with the argument of the particular attachment of the ancient stoics to military metaphors (Stankiewicz, 2014, p. 227). That is a typical argument of *ipse dixit* type.

Secondly, he left his two additional theses (the additional thesis II and the additional thesis III) without substantiation.

He did not explain (additional thesis II) why we should "visualize concretely and in detail" our potential misfortunes (Stankiewicz, 2014, p. 224). That thesis is only supported by the subthesis of the additional thesis II which says: "the more we force our imagination the more effectively we will premeditate the misfortune" (Stankiewicz, 2014, p. 224). Once more, that subthesis is left without substantiation.

The thesis about the chess-player's strategy is also not supported by any argumentation (*additional thesis III*) unless – following the rule of kindness (see Kisielewicz, 2014, pp. 174–176) – I accept that it is a particular case of PM and should be added to Thesis I, which then would be more complete:

Thesis I completed: We have to foresee or plan (like a chess-player) each evil event or a worse variant of a possible situation (for example in a situation "I win–I lose" I chose the second variant).

P. Stankiewicz also did not manage to avoid contradictions between *additional thesis I* (we do not take into consideration misfortunes of little possibility) and the *argument overthrowing anti-thesis IV* (remember about misfortunes of little possibility). The contradiction consists of postulates which preclude one another and refer to implausible events (of little plausibility). On the one hand, he postulates to not include such events in PM for two reasons

– it is hard to foresee them (the first argument supporting additional thesis I) (Stankiewicz, 2014, p. 223),
– they do not start the mechanism of PM (the first argument supporting additional thesis I) (Stankiewicz, 2014, p. 225),

and at the same time he states that you mustn't forget about misfortunes of little plausibility (argument overthrowing anti-thesis IV) so despite of everything –also in spite of the above two reasons! – these improbable (of little probability) events should be included in the PM because it is better than being and incurable optimist (Stankiewicz, 2014, pp. 234–235).

The remaining argumentation by P. Stankiewicz is to the point and adequate, although in one place we have to refer to the principle of the charity again (Kisielewicz, 2014, pp. 174–176). In the case of an argument overthrowing anti-thesis I, we have to trust the author that he characterized the differences between a hopeless doomsayer and someone practicing PM accurately (Stankiewicz, 2014, p. 230). We must believe him that his synthesizing sentences are based on solid knowledge.

The rest of the argumentation is in my opinion convincing. The strongest moment is the third argument supporting Thesis I which is the argument of possible surprise which is always positive (Stankiewicz, 2014, p. 226).

The philosophical narration of Piotr Stankiewicz is of literary–persuasive nature typical for the genre to which his book belongs. It is written in an approachable way and rhetorical and stylistic elements lead us closer to the author's objective. In that sense, the text of *The Art of Living According to Stoics* [Sztuka życia według stoików] is praiseworthy for its simplicity which means it is "easily understandable".

Nevertheless, there are logical costs of that simplicity. The reader has to decide on his/her own whether in the guidebook to a "good living" he or she looks for impeccable logical argumentation or for an imaginative disquisition that is not logically coherent at every point. In other words, the reader is facing a dilemma: do I build – to quote Pierre Hadot – an "internal fortress", resistant to logical hits of adversaries or to follow the beauty of persuasion at the cost of leaving gaps in that fortress.

Now I will present an analytical way of substantiating PM. The adjective analytical I understand here as "aiming at clarity, precision and rigorous argumentation" (The Philosophical Gourmet Report 2000–2001; in: Prechtl 2009, p. 11).

3 Analytical Substantiation of PM

Thesis: PM (*praemeditatio malorum*) is a better strategy than PB (*praemeditatio bonorum*).

3.1 I adapt the set of four mental events (=ME) for this study. That set includes: imagined suffering (=IS), suffering which is being currently experienced (in short: current suffering =CS), imagined pleasure (=IP) and pleasure which is being currently experienced (in short: current pleasure =CP), therefore ME = {IS, IP, CS, CP}.

3.2 The ME set will be arranged intuitively and naturally (terminology after Kraszewski, 2012, p. 106). It is possible because there is an adequate relation ordering each pair of elements, the relation of being preferred by mentally healthy people. From each pair of elements of ME, one element will be preferred by mentally healthy people at the cost of the other. The elements of the ME set create six pairs: IS and IP, IS and CS, IS and CP, IP and CS, IP and CP, CS and CP.

Statement 3.2: A mentally healthy person when given one element to choose instead of the other prefers (chooses): IP from the pair IS and IP, IS from IS and CS, CP from IS and CP, IP from IP and CS, CP from IP and CP as well as CP from CS and CP.

Substantiation 3.2

a. A mentally healthy person – having to choose to imagine a toothache (IS) or to imagine resting under the palm trees on an exotic island (IP) – will prefer IP over IS.
b. A mentally healthy person – having to choose to imagine a toothache (IS) or an actual tooth-ache (CS) – will prefer IS over CS.
c. A mentally healthy person – having to choose to imagine a toothache (IS) or an actual resting under the palm trees on an exotic island (CP) – will prefer CP over IS.

d. A mentally healthy person – having to choose an actual toothache (CS) or to imagine resting under the palm trees on an exotic island (IP) – will prefer IP over CS.

e. A mentally healthy person – having to choose to imagine resting under the palm trees on an exotic island (IP) or actually resting under the palm trees on an exotic island (CP) – will prefer CP over IP.

f. A mentally healthy person – having to choose an actual toothache (CS) or actual resting under the palm trees on an exotic island (CP) – will prefer CP over CS.

3.3 We have six possible relations out of which CP is preferred three times, IP – is preferred twice, IS – is preferred once, CS is not preferred at all. It is shown in the following recording of the relation P(x, y) "x is preferred over y" (where the first argument is the preferred element – which is the 'winner' of that relation):

P(CP,IS)
P(CP, IP)
P(CP, CS)
P(IP, IS)
P(IP, CS)
P(IS, CS)

3.4 Depending on how many times a given element was preferred we ascribe it a numerical value:

CP = 3
IP = 2
IS = 1
CS = 0

3.4.1 In accordance with the theory of the empirical meaning of a sentence by A. Kisielewicz (2014, p. 119) most users of the English language will agree that a mental event of higher value is better than a mental event of a lower value. Therefore most speakers of English will agree that CP is better than IP, IS and CS, IP is better than IS and CS, and IS is better than CS (see also Examples 3.2). If the two-element predicate B (a, b) "a is better than b" we present as the relation of aLb, we will receive a multiple relations

(CP) B(IP) B(IS) B(CS)

isomorphic to

3>2>1>0.

3.5 The ME set ordered by the relation aLb (see 3.3–3.4.1) is a sequence

(CS, IS, IP, CP)

or – after we substitute numerical values of the mental events from 3.4 – a number sequence

(0, 1, 2, 3).

From now on instead of "mental event of the value of 0, 1, 2 and 3" I will say "0, 1, 2, 3" respectively, also, instead of "CS, IS, IP, CP" I will say"0, 1, 2, 3" respectively.

3.6 These agreements allow for an analytical definition of PM and PB

Definition 3.6.1

(α) PM is a sequence of two different mental events which:

(β) do not take place at the same time, one of them takes place in t_1 time, and the other – in t_2 time,

the first of them is 1.

Example 3.6.1

PM, the imagined premeditation of misfortune, assumes an earlier occurrence of 1. After 1 three different mental events may occur: 0, 2 and 3. For example I first imagine a toothache (1), after the act of imagination the following take place: either actual (not imagined) suffering (0), or a more pleasant act of imagination (2), or an actual (not imagined) pleasure (3).

Definition 3.6.2

PB is a sequence of two different mental events which:

a. do not take place at the same time, one of them takes place in t_1 time, and the other – in t_2 time,

b. the first of them is 2.

Example 3.6.2

PB, the imagined premeditation of fortunate events, assumes an earlier occurrence of 2. After 2 three different mental events may occur: 0, 1, and 3. For example I first imagine resting under palm trees (2), after that act of imagination the following takes place: either actual (not imagined) suffering for example a toothache (0), or an unpleasant act of imagination (1), or an actual (not imagined) pleasure (3).

3.6.1 The PM strategy – in accordance with definition 3.6.1 and example 3.6.1 – may be presented in the form of three different functions:

	$f_1(x)$		$f_2(x)$		$f_3(x)$	
x	t_1	t_2	t_1	t_2	t_1	t_2
y	1	0	1	2	1	3

3.6.2 The PB strategy – in accordance with definition 3.6.2 and example 3.6.2 – may be presented in the form of three different functions:

	$f_4(x)$		$f_5(x)$		$f_6(x)$	
x	t_1	t_2	t_1	t_2	t_1	t_2
y	2	0	2	1	2	3

3.7 Amongst the six different functions $f_1(x)$-$f_6(x)$ from 3.6.1–3.6.2 three of them are increasing functions (where along with the increase of the argument the function increases) and three decreasing functions (where along with the decrease of the argument the function decreases). The increasing functions are:

	$f_2(x)$		$f_3(x)$		$f_6(x)$	
x	t_1	t_2	t_1	t_2	t_1	t_2
y	1	2	1	3	2	3

and the decreasing functions are:

	$f_1(x)$		$f_4(x)$		$f_5(x)$	
x	t_1	t_2	t_1	t_2	t_1	t_2
y	1	0	2	0	2	1

Definition 3.7.1

The increment of the function value in accordance with 3.4.1 we call *mental improvement*, and the decrement of the function value in accordance with 3.4.1 we call *mental deterioration.*

Definition 3.7.2

The value of mental improvement and mental deterioration is (symbol "↑" and "↓" respectively) the difference of the higher function value and the lower function value.

3.8 After adapting these definitions we state that mental improvement of the total value of 4 occurred in:

	$f_2(x)$		$f_3(x)$		$f_6(x)$	
x	t_1	t_2	t_1	t_2	t_1	t_2
y	1	2	1	3	2	3
Value↑	1		2		1	
Total			1+2+1=4			

whereas deterioration of a mental event of a total value of 4 occurred in:

	$f_1(x)$		$f_4(x)$		$f_5(x)$	
x	t_1	t_2	t_1	t_2	t_1	t_2
y	1	0	2	0	2	1
Value↓	1		2		1	
Total			1+2+1=4			

3.9 In accordance with definition 3.6.1 and definition 3.6.2 we substitute $f_n(x)$ in tables from 3.8 with PM and PB and we receive a table of improvements

	PM		PM		PB	
x	t_1	t_2	t_1	t_2	t_1	t_2
y	1	2	1	3	2	3
Value ↑	1		2		1	
Total	1+2=3				1	

and a table of deteriorations

	PM		PB		PB	
x	t_1	t_2	t_1	t_2	t_1	t_2
y	1	0	2	0	2	1
Value ↓	1		2		1	
Total	1		3			

which can be presented in this graph

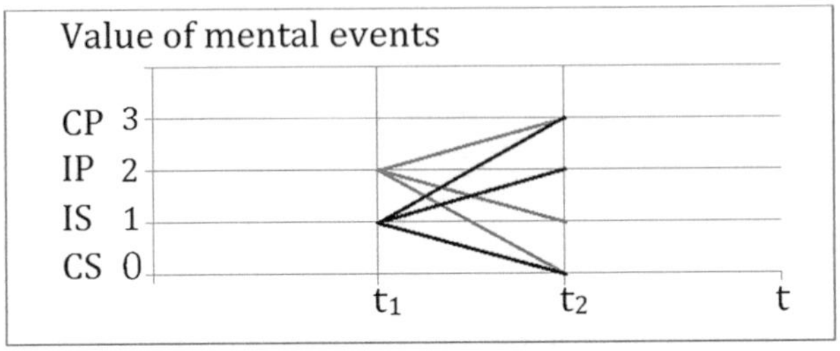

Conclusion. PM is better than PB, because

a. PM (color black) includes two improvements of a total value of 3, whereas PB (color grey) – one improvement of a total value of 1,

b. PM (color black) includes one deterioration of a total value of 1, whereas PB (color grey) has two deteriorations of a total value of 3.

Quod erat demonstrandum

4 Conclusions

In this article, I presented two philosophical narrations – the traditional one (persuasive-literary) and the analytical one. I correlated them in search for the answer to the question: what are the costs of simplicity and complexity (complications) of philosophical discourse? Does a simple philosophical text (easily understandable, full of rhetoric tricks and literary panache) have to lose the coherence of logical argumentation? Is a complex philosophical text (complicated, difficult to understand, requiring effort during reading, full of internal references) deprived of literary charm? These are the conclusions.

The literary persuasion of P. Stankiewicz is fuller than my analysis. The philosopher presents more aspects of PM while I only correlated PM with PB and I drew a conclusion from that correlation that PM is superior to its opponent PB. The argumentation of P. Stankiewicz might have logical gaps here and there, but is beautiful in its literariness, what the readers of his book may see for themselves (Stankiewicz, 2014). My explanations have no claims towards the literary panache but the argumentation coherence is missing – I hope – larger logical breaches. However the essential question is inevitable here: does a dry, quasi-mathematical style, deprived of ornaments and rhetorical power, have the capability to persuade a person who is looking for a piece of spiritual advice, for the answer to the question about decent and quiet life?

I am not able to solve that query. While not hiding my analytical leaning, I leave that task to the reader.

References

Becker, L. C. (1999). *A new stoicism*. Princeton, NJ: Princeton University Press.

Fabjański, M. (2010). *Street stoicism. How to tame difficult situations* [*Stoicyzm uliczny. Jak oswajać trudne sytuacje*]. Warszawa: Czarna Owca.

Fabjański, M. (2014). *Trust life. Do not fall in love with a sparrow flying By* [*Zaufaj życiu. Nie zakochuj się w przelatującym wróblu*]. Warszawa: PWN.

Hadot, P. (2003). *Philosophy as a way of life: Spiritual exercises from Socrates to Foucault* [*Filozofia jako ćwiczenie duchowe*] (P. Domański, Trans.). Warszawa: Aletheia.

Irvine, W. B. (2009). *A guide to the good life. The ancient art of stoic joy*. New York, NY: Oxford University Press.

Irvine, W. B. (2013). *A slap in the face. Why insults hurt, and why they shouldn't*. New York, NY: Oxford University Press.

Kisielewicz, A. (2014). *Logic and common sense. New approach to logic and argumentation* [*Logika i zdrowy rozsądek. Nowe podejście do logiki i argumentacji*]. Wrocław: (unpublished book, kindly provided to me by the Author in a PDF format).

Kraszewski, J. (2012). *Introduction to mathematics* [*Wstęp do matematyki*]. Warszawa: WNT.

Mazur, T. (2012). *Disaster. A handbook of failed existence* [*Fiasko. Podręcznik nieudanej egzystencji*]. Warszawa: Cartablanca.

Mazur, T. (2014). *On becoming a stoic. Are you ready for success* [*O stawaniu się stoikiem. Czy jesteście gotowi na sukces*]. Warszawa: PWN.

Prechtl, P. (2009). *Lexicon of analytical philosophy terms* [*Leksykon pojęć filozofii analitycznej*] (J. Bremer, Trans.). Kraków: WAM.

Reale, G. (2010). *A history of ancient philosophy III. Systems of the Hellenistic age* [*Historia filozofii starożytnej. III. Systemy epoki hellenistycznej*]. (E. I. Zieliński, Trans., 3rd ed.). Lublin: Wydawnictwo KUL.

Seneca, L. A. (1961). *Moral letters to Lucilius* [*Listy moralne do Lucyliusza*] (W. Kornatowski, Trans, introduction and notes K. Leśniak). Warszawa: PWN (Biblioteka Klasyków Filozofii).

Seneka, L. A. (1989). *Dialogs* [*Dialogi*] (L. Joachimowicz, Trans.). Warszawa: Pax.

Stankiewicz, P. (2014). *The art of living according to stoics. How to live wisely, well and happily* [*Sztuka życia według stoików. Jak żyć mądrze, dobrze i szczęśliwie*]. Warszawa: WAB.

CHAPTER 13

The Complexity and Simplicity of Zinc Proteins

Minimum Form, Maximum Substance

Artur Krężel

Abstract

The proteins present in our bodies are not only properly structured polypeptide chains composed of amino acids, but also macromolecules binding a variety of substances altering their stability and function. These substances, of which the most popular are metal ions, including zinc ions, are called cofactors. It is believed that as many as 10% of the proteins in our body (approx. 3,000) are zinc proteins, which form structures and functions with a distinctive complexity. Despite such a vast diversity of zinc proteins, their common denominator is the zinc ion, which in all of them is bound by only the four amino acid residues in virtually the same, or a very similar, geometry reflecting the simplicity and minimalism of these complex molecules. What, then, is responsible for the diversification of functions, the stability, and the many physicochemical and biochemical characteristics of zinc proteins? I will answer this question, presenting three types of zinc proteins: metalloenzymes, proteins in which the zinc ion performs functions of a structural kind, and protein complexes, in which the zinc ion is responsible for the spatial organization of the complex. The article describes the factors affecting the differentiation of the kinetic and thermodynamic stability of the zinc complex in proteins which, in consequence, leads to the enormous complexity of these macromolecules in respect of their biological function. The aim is to illustrate how nature, using the same simple molecular building blocks, creates something that, in its form and function, is complex: namely, the so-called zinc metalloproteome.

Keywords

protein – zinc – stability – sequence – structure

1 Introduction

Proteins are large molecules formed of building blocks called amino acids. Classical biochemistry commonly says that they are composed of 20 different amino acids, differing in structure from each other; however, currently, this statement is not considered so unequivocally. In addition to the amino acid building blocks, proteins are composed, among others, of the so-called cofactors, i.e., chemical substances other than amino acids, directly bound with proteins that modulate their function. Proteins utilize a large number of different cofactors so that to achieve various structures and properties. One of the most important of these substances are metal ions, which, together with the so-called macromolecules, form metalloproteins, such as metalloenzymes, structural proteins containing metal cores, proteins accumulating metal ions and transporting them. Of all the metal cofactors, transition metal ions play a unique role in the differentiation of function and structure of metalloproteins (Maret & Li, 2009). Among them, the most abundant metal ion in all domains of life is a zinc ion (formally Zn^{2+}), which forms with proteins the enormity of structures serving various functions (forming the so-called zinc metalloproteome). Previous studies as well as bioinformatic analysis of the human genome indicate that as many as 10% of the proteins encoded in the genome are zinc proteins (Andreini et al., 2006). This in turn converted into numerical values constituting 3,000 various biomolecules (*complexity*). Bioinformatic analysis is largely based on sequence homology (signatures), and does not take into account interactions between proteins, which in consequence may mean that there are significantly more zinc proteins than we currently believe (Maret, 2006).

Proteins binding Zn^{2+} ions differ significantly from each other in size, spatial structure, organization into larger complexes or eventually biological function. Despite these differences, the Zn^{2+} ion is bound by proteins using only four amino acid residues, predominantly in the tetrahedral geometry (*simplicity*). These include cysteine (C), histidine (H), aspartic (D) and glutamic acids (E) residues. The diversity of zinc protein function is not based on a simple combination of these amino acid residues, as this would not lead to the formation of thousands of different molecular forms. This work describes how the zinc enzymes and proteins with zinc ion structural binding sites are modulated by a series of factors leading to the enormous diversity of their function and stability.

2 Modulating Metalloproteins' Properties via Ligands

Metalloproteins, through specific atom combinations (ligands) involved in the formation of the catalytic or structural site, are able in some way to modulate both their properties and affinity for Zn^{2+}. The already classic Pearson's theory of hard and soft acids says that proteins containing sulfur and nitrogen ligands are substantially more stable than those containing Zn^{2+} ion bound by nitrogen and oxygen ligands (Pearson, 1963). However, a comparison of literature data on the stability of proteins significantly different in terms of binding mode, e.g., alkaline phosphatase (DDHx, DHHx), metallothionein (CCCC) or zinc fingers (CCHH) demonstrates a very similar affinity for Zn^{2+} ions, within the picomolar range (10^{-11}–10^{-12} M). Although the stability of cellular zinc proteins is similar, the thermodynamics of Zn^{2+} binding differs between individual objects (Kochańczyk et al., 2015). Studies conducted on zinc motives showing variable coordination of CCHH, CCHC and CCCC types, with similar affinities, showed different entropic and enthalpic contributions to the free enthalpy of the system (stability). For example, the binding of Zn^{2+} to the CCC motif is driven mainly entropically, which makes this reaction more susceptible to changes in factors such as pH or temperature. With fewer sulfur donors, entropy becomes less favored and the overall stability of the system depends mainly on the enthalpic component (Rich et al., 2012).

On the other hand, proteins containing identical arrangements of ligands associated with Zn^{2+}, e.g., HHEx (where x represents a variable external ligand) may exhibit very different affinities for, e.g., prolactin (10^{-5} M), angiotensin-converting enzyme inhibitors (10^{-8} M) and thermolysin (10^{-11} M). This proves that the so-called first sphere of coordination (Zn^{2+} binding site) is not the main factor affecting the stability of zinc protein. This means that the so-called further protein regions play a key role in the stabilization or destabilization of macromolecules through the interaction of non-binding amino acid residues with each other. Carbonic anhydrase is an example of an extremely thermodynamically and kinetically stable protein, in which the Zn^{2+} ion is bound by three histidine residues (HHHx). Its stability is enhanced by the simultaneous action of ligands binding Zn^{2+} with other amino acid residues in the further regions of the protein through the formation of hydrogen bonds. Mutation of residues involved in the formation of hydrogen bond network, not directly involved in the coordination of Zn^{2+} ions, causes a decrease in the thermodynamic and kinetic stability of the protein (Maret & Li, 2009). Another interesting example of secondary interactions affecting the stability of macromolecules is associated with the simultaneous interaction of ligands binding Zn^{2+} with a peptide bond via hydrogen bonds (-NH⋯S-). One example is the HIV-1 nucleocapsid protein, NCp7 (Namuswe & Berg, 2012).

3 Structural Diversity and Simplicity of Zinc Finger Proteins

The so-called classical zinc fingers show unusual characteristics of structural diversity and at the same time the simplicity of Zn^{2+} binding method. There have been thousands of them identified in different organisms so far, and they are the second class of protein domains in terms of abundance. Despite such a great number, they show high sequence similarity – they contain only approx. 30 amino acids arranged in a highly canonical sequence. The classical zinc finger sequence is as follows: (F/Y)-X-C-$X_{2\text{-}4}$-C-X_3-(F/Y)-X_5-L-X_2-H-$X_{3\text{-}5}$-H, where X, M, Y and L represent any amino acid, phenylalanine residue, tyrosine and leucine, respectively (Miłoch & Krężel, 2014). Zn^{2+} ion binds to this motif via two cysteine and histidine residues, forming a characteristic ββα structure responsible in most cases for specific binding to the correct DNA sequence. The metal binding method is identical in thousands of known zinc finger proteins, which has been proven by numerous structural studies. One can safely say that the simplicity of Zn^{2+} ion binding is manifested in the diversity of forms playing a similar function, although specialized towards different molecular targets. Removal of any amino acid responsible for DNA identification in a zinc finger loop results in a very marked decline in the stability of the entire system. Similarly, interference with the so-called hydrophobic cortex of a finger through mutations to less hydrophobic amino acids results in a decrease in stability and a tendency to form other, non-functional structures. Substitution of any donor in a classical zinc finger, e.g., histidine residue to aspartic acid residue, causes dramatic changes in the structure and affinity for Zn^{2+} ion, which is reflected in the functional binding of zinc fingers to DNA (Kochańczyk et al., 2015).

Another relationship between the simplicity and complexity of zinc proteins is exemplified in the presented above stability of zinc domains. The same mode of Zn^{2+} ion binding in the zinc finger domain and a highly conserved sequence would suggest almost identical stability. It turns out that natural zinc fingers are quite diverse in terms of stability, which is perfectly reflected in the so-called consensus zinc finger, being the most representative sequence. The consensus zinc finger contains amino acid residues that occur most frequently at a given position. Its stability is higher than that of natural fingers by more than 1,000 times (Sénèque & Latour, 2010). This is mainly due to a series of interactions occurring in the consensus zinc finger itself, which stabilize the whole structure. Natural zinc fingers do not have that many stabilizing effects within a single structure, which causes a reduction in their stability. Furthermore, the distance between the individual amino acid residues in a sequence of zinc finger also has a significant effect on the modulation of the Zn^{2+} ion affinity. In conclusion, it can be seen that the factually identical metal ion binding mode in a very similar spatial structure (sequence and structure simplicity), as a consequence, manifests itself in different stability and function (stability and function diversity) (Kochańczyk et al., 2015).

4 Complexities of Interprotein Zinc Binding Sites

Although many key zinc proteins have not yet been characterized in terms of affinity for Zn^{2+}, definitely the interprotein binding sites of this metal remain the most enigmatic and the least understood. These sites are formed by two or more protein molecules, which bind Zn^{2+} ions together. They both form homo- and heteromer-type structures. Isolation of this type of complexes as well as the characteristics of interprotein binding sites is extremely difficult and available bioinformatic tools are still insufficient to predict them. Isolation of complexes from natural sources or organisms for their overproduction usually results in proteins lacking Zn^{2+} or individual protein units not assembled in higher structures by the metal ion. An example of a protein, in which this type of interaction occurs is RAD50, a part of a highly conserved DNA repair complex – MRE11. This protein forms a homodimer, in which the Zn^{2+} ion is bound by four cysteine residues, each two derived from two identical RAD50 units, to form the so-called hook zinc domain (Stracker & Petrini, 2011). Although the binding of Zn^{2+} by cysteine residues is favored entropically, the formation of a complex of Zn-(protein)$_2$ stoichiometry becomes unfavorable. There must be other mechanisms present, determining the sufficiently high stability of this site, in order to form this type of complex under physiological conditions. Studies by Kochańczyk et al., 2013 have shown that the formation of an intermolecular hydrophobic network and the formation of hydrogen bonds within peptide chains of the Zn^{2+} binding motif are both responsible for this stabilization (Kochańczyk et al., 2013). Interesting, in terms of complexity and simplicity, is the fact that the zinc hook motif is present in all domains of the living world. It occurs also in the viral proteins. It is not the simplicity of binding of two protein molecules through the coordination of zinc ions that makes it so extremely important, but the manner of maintaining complex stability in organisms, in which Zn^{2+} ion metabolism is highly diversified. As it appears, these are the sequence and structure subtleties (e.g., variations in one or more amino acid residues) that decide whether the protein is stable enough to function, e.g., in thermophilic archeon, or moderately stable as in eukaryotes, so that the kinetics of dissociation and association of Zn^{2+} ions could occur at high speed necessary for the proper functioning of the MRE11 complex.

5 Conclusions

In summary, metalloproteins, and zinc proteins in this particular case, acquire their numerous structures and functions through the binding of small Zn^{2+}

ions. The metal-binding method is simple and common to many proteins differing in structure and biological function. Diversification of properties, particularly the thermodynamic stability discussed here, is influenced not only by the bound ligands but also by their additional interactions occurring in other protein regions.

References

Andreini, C., Banci, L., Bertini, I., & Rosato, A. (2006). Zinc through the three domains of life. *Journal of Proteome Research*, *5*(11), 3173–3178.

Kochańczyk, T., Drozd, A., & Krężel, A. (2015). Relationship between the architecture of zinc coordination and zinc binding affinity in proteins–insights into zinc regulation. *Metallomics: Integrated Biometal Science*, *7*(2), 244–257.

Kochańczyk, T., Jakimowicz, P., & Krężel, A. (2013). Femtomolar Zn(II) affinity of minimal zinc hook peptides – a promising small tag for protein engineering. *Chemical Communications*, *49*(13), 1312–1314.

Maret, W. (2006, April 15). Protein interface zinc sites: The role of zinc in the supramolecular assembly of proteins and in transient protein-protein interactions. In *Handbook of metalloproteins*. John Wiley & Sons, Ltd. https://doi.org/10.1002/0470028637.met016

Maret, W., & Li, Y. (2009). Coordination dynamics of zinc in proteins. *Chemical Reviews*, *109*(10), 4682–4707.

Miłoch, A., & Krężel, A. (2014). Metal binding properties of the zinc finger metallome–insights into variations in stability. *Metallomics: Integrated Biometal Science*, *6*(11), 2015–2024.

Namuswe, F., & Berg, J. M. (2012). Secondary interactions involving zinc-bound ligands: Roles in structural stabilization and macromolecular interactions. *Journal of Inorganic Biochemistry*, *111*, 146–149.

Pearson, R. G. (1963). Hard and soft acids and bases. *Journal of the American Chemical Society*, *85*(22), 3533–3539.

Rich, A. M., Bombarda, E., Schenk, A. D., Lee, P. E., Cox, E. H., Spuches, A. M., Hudson, L. D., Kieffer, B., & Wilcox, D. E. (2012). Thermodynamics of Zn2+ binding to Cys2His2 and Cys2HisCys zinc fingers and a Cys4 transcription factor site. *Journal of the American Chemical Society*, *134*(25), 10405–10418.

Sénèque, O., & Latour, J.-M. (2010). Coordination properties of zinc finger peptides revisited: Ligand competition studies reveal higher affinities for zinc and cobalt. *Journal of the American Chemical Society*, *132*(50), 17760–17774.

Stracker, T. H., & Petrini, J. H. J. (2011). The MRE11 complex: Starting from the ends. *Nature Reviews. Molecular Cell Biology*, *12*(2), 90–103.

CHAPTER 14

Why Do We Use Simple Organisms to Model Complex Human Diseases?

Marta Migocka-Patrzałek

Abstract

A model organism is a species, different from the human one, which we investigate intensively to gain insights about specific biological processes, hoping to grasp the more widely applicable, general laws that regulate the functioning of organisms. Animal models are also used to more comprehensively understand human diseases. Such a strategy is efficient given the common origins of life on Earth and with this the shared developmental and metabolic pathways that owe their existence to the retention of features of the genetic code throughout evolution. One can, therefore, certainly ask why it is that we make use of simple organisms, such as worms, fruit flies, or fish, in our research? Can they really give us answers regarding complex human diseases? At the very least, it is surely worth knowing more about how such research is designed and conducted to simply evaluate its potential benefits.

Keywords

biology – model organisms – human diseases

1 Introduction

Animals have been used not only to study the basic principles of life, but also to better understand the anatomy, and physiology of living organisms. Moreover, animals play a central role in education, pharmacology and research. The ability to conduct experiments in controlled situations, and to mimic the biological conditions of human disease in animals led to the development of scientific methods and originated in the concept of animal models of human diseases. The model organism is a species, different from humans, which we intensively investigate to gain knowledge about specific biological processes and to understand wider, general laws that rule over organisms functioning. Although the use of animals in research has been questioned by the scientific community due to ethical concerns, the foundations of our basic knowledge of human physiology and anatomy are based on preclinical research with the use of animals. This is confirmed by the fact that nearly 90% of research awarded with the Nobel Prize in Physiology and Medicine, was conducted using animals in experimental investigations. It is also important to note that animal studies allow researchers to carry out experiments that would be impractical or ethically prohibited with humans (Andersen & Winter, 2019; Burggren & Warburton, 2007; Dubińska-Magiera et al., 2022; Plantié et al., 2015).

2 Why Does a Particular Animal Become a Research Model?

The main reason is availability and the ease of maintenance. For example, mice, chickens, frogs and fruit flies (*Drosophila melanogaster*) (Figure 14.1) are widely used just because they are easily accessible and simply cheap. The anecdote regarding one of the frog species, called the African clawed frog (*Xenopus*), is a nice example showing why a particular animal becomes a research model. Once upon a time, scientists discovered that the urine of pregnant women, containing hormones, can induce the development of *Xenopus* ovaries. Therefore, between the 1940s and 1960s thousands of frogs started to act as living pregnancy tests all around the world. The *Xenopus,* abundant in laboratories and easy to maintain, can be utilized in other kinds of experiments, trials and research, and as a consequence became a useful animal model. In general, the amphibians turned out to be great models in a variety of research and took a profound place in scientific areas such physiology of musculoskeletal, cardiovascular, renal, respiratory, reproductive, and sensory systems (Burggren & Warburton, 2007; Kiełbwna & Migocka-Patrzałek, 2017).

To be used as a model, the species must meet certain criteria that enable the feasibility of experimental procedures. From the researcher's point of view, the high rate of reproductivity and relatively short life cycle (the time between being born and the ability to reproduce) is a very valuable feature. In the case of mice, having an accelerated lifespan, we can assume that one mouse year equals about 30 human years. Therefore, their whole life cycle could be examined within only two or three years. It is beneficial, especially in genetic research, in which crossing individuals to obtain several, subsequent generations in a relatively short time, is crucial. The financial feasibility and previous history of research with the use of a particular species are also useful. Such experimental data allows for research outcome comparison with the other

FIGURE 14.1 Fruit fly (*Drosophila melanogaster*). Fruit flies are one of the most popular model organisms. With the low cost of husbandry, short generation time, and variety of molecular tools for genetic engineering, flies become a useful model in biomedical experiments

researchers' results, therefore not only enabling but also greatly facilitating the conclusions drawing process. The known genetic code of animals is crucial for genetic research since the possibility of genetic manipulation is nowadays one of the most important research techniques.

Choosing an animal model for a particular purpose is a complex decision, involving scientific and practical considerations. Many species are used in biomedical research, such as yeasts, insects, nematodes (*Caenorhabditis elegans*), fish (*Danio rerio* – zebrafish), frogs and many mammals, such as mice, rats, dogs, cats, pigs and even monkeys. Several other issues should be also taken into account during experiment planning, such as the availability of adequate techniques and methods, including equipment, and molecular tools (such as specific antibodies, etc.) (Andersen & Winter, 2019; Dubińska-Magiera et al., 2016; Ericsson et al., 2013; Robinson et al., 2019).

3 Animal Models of Disease Can Be Divided into Two Categories, Spontaneous and Induced

The use of animals as a model of human diseases had increased dramatically in the twentieth century. The beginnings of detailed biological analysis of particular disorders were possible because animals sometimes suffer from the same kind of diseases as humans. Also, naturally occurring animal genetic variants (mutants) can be utilized as a model of a particular human genetic disease. We call such animals a spontaneous model. There are hundreds of well-characterised animal strains with inherited genetic disorders, allowing us to examine in detail human genetic diseases (see, e.g., the Jackson Laboratory database[1]). Mice, called sometimes "lifesavers", because of their usefulness in medical trials, are biologically very similar to humans and get many of the same diseases, for the same genetic reasons. One such example is the so-called, nude mice. Its genome contains a spontaneous mutation in the gene *Foxn1nu*, leading to two main defects: abnormal hair growth and defective development of the thymic epithelium. As a result, the nude mice do not develop thymus, therefore they lack the T cells, which are crucially important for natural, immunological defence. This model is widely used to induce diverse kinds of cancer by injecting cancer cell lines into nude mice, enabling preclinical testing of anticancer drugs (Cordier & Haumont, 1980; Onaciu et al., 2020; Szadvari et al., 2016).

On the other hand, nowadays, we have learned how to induce particular pathological conditions in animals to create disease models. Such induction could be a combination of physical and chemical factors, or targeting genetic manipulation. The induced models are based on healthy animals, in which the pathological disorder to be examined is experimentally induced. Examples include partial hepatectomy to study liver regeneration, or studies of cow's milk allergy in mice by immunizing them with tiny doses of cow milk protein (reviewed in: Hau, 2008). Animals can also carry out a mutation in certain, important genes and suffer from genetic diseases, similar to humans. A genetic disease is a disorder caused by an abnormality in an individual's genome. The genetic disease may be caused by a mutation in a single gene but can be also caused by different factors such as a combination of environmental factors and mutations in multiple genes. The genetic disease could come also from chromosome alterations or mutations in the non-chromosomal DNA of mitochondria (Dubińska-Magiera et al., 2016; Plantié et al., 2015). Mutants are extremely rare in nature, therefore, the acceleration of research conducted on animal models of human diseases took place when mutagenesis techniques became available. Effective mutagenesis tools were needed to make genetic approaches functional and practical. The fact that X-rays could induce mutations in mice was known from 1923. Then in 1980 extensive, chemical mutagenesis studies were continued using the alkylating chemical called ENU (N-ethyl-N-nitrosourea). Next, in 1990, the human genome sequencing project began, aiming for the identification and mapping of all of the genes in the human genome. The project progress was unexpectedly fast and created an urgent need for knowledge regarding particular gene's functions. Therefore, approximately in the same period, the idea of a large-scale mouse mutagenesis

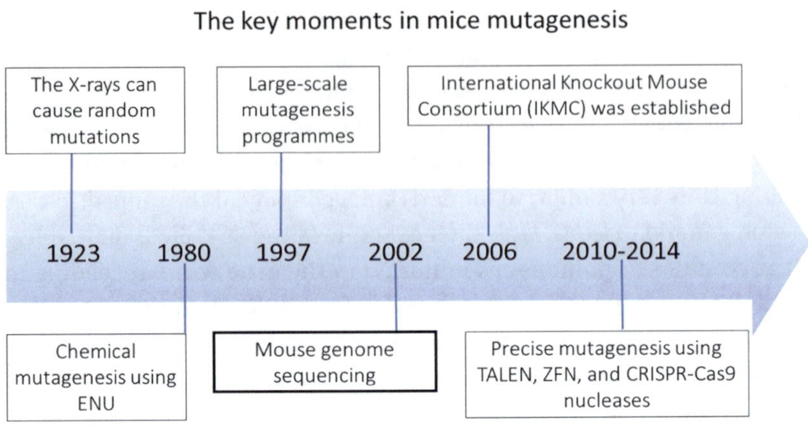

FIGURE 14.2 The key moments in mice mutagenesis. The aim of mutagenesis is to establish an animal model of human genetic diseases. The progress in genetic engineering starts from the discovery that X-rays can induce mutagenesis. Then the chemical compound ENU (N-ethyl-N-nitrosourea) was used to create mutations. With the technique development, many large-scale programs started in 1997, with the culmination point in 2007, when the International Knockout Mouse Consortium (IKMC) was established, with the aim to provide knockout and conditional mouse strains for all mouse genes. Such acceleration was possible because the whole mouse genome was sequenced in 2022. Starting from 2010 new techniques, allowing for precise mutagenesis were discovered. In the beginning, the ZNFs and TALENS were developed. Then, the CRISPR-Cas9 technique was shown to be a real scientific breakthrough, appreciated by the Nobel Prize awarded in 2020 to Emmanuelle Charpentier and Jennifer Doudna (according to: Chenouard et al., 2021; Gondo et al., 2009)

project emerged. The approach is based on large-scale, random mutagenesis using ENU. Then the changes in phenotype, such as disease symptoms, were observed. Finally, the genome of selected individuals was sequenced to find out which gene is responsible for a particular disorder. The similarity between mice and human genes is high enough to let us extrapolate the observations to use mice in diagnosis and medicine. Further progress in molecular methodology and molecular tools for directed mutagenesis (allowing for specific, directed mutation in a particular gene) led to the establishment of the International Mouse Phenotyping Consortium (IMPC) in 2006. Its goal is to fill the gaps in the knowledge by use of the reverse genetics approach. This procedure is called gen-driven mutagenesis, where the target gene is disrupted, and the results are drawn through the phenotypic analysis (reviewed in: Chenouard et al., 2021; Gondo et al., 2009) (Figure 14.2).

The way from primary, spontaneous animal models of human diseases to directed mutagenesis, allowing for the customization of genetic modifications, was difficult but successful. However, we should keep in mind that the biological interaction between genes and other biological factors is a very complex and multidimensional process. Therefore, detailed studies are needed to understand the whole picture of the genetic disease pathological process.

4 The Efforts toward Understanding the Disease's Genetic Source

We, humans, share the vast majority of genes with animals such as mice, zebrafish, and even fruit flies. The similarities between human and mouse genomes allow us to assume that whenever candidate genes for human diseases are found, their homologs usually are present in, e.g., the mouse genome. On the other hand, we can mutate genes with unknown functions and observe the phenotype. Such an approach guides us to the discovery of genes, which disruption can lead to diseases. As an example, we can mention the research results published by (Dickinson et al., 2016) regarding the high-throughput discovery of novel developmental phenotypes. The scientists identified some disorders and changes in phenotypes in different developmental stages, which were the results of mutations in many previously uncharacterized genes (Groza et al., 2023; Muñoz-Fuentes et al., 2018). The effort of several international projects aiming to investigate the function of mouse genes by mutagenesis, and following molecular analysis and phenotyping, greatly expand our knowledge. This approach creates the possibility of using the mouse as a valuable organism to interpret the information coded in the human genome. For example, the phenotypic analysis shows that mutation in the transmembrane protein 132A gene, *Tmem132a*, leads, among others, to abnormalities in the animal spinal cord structure and narrower limbs (Rosenthal & Brown, 2007). Such information could make the identification of genetic disease sources easier. Also, can help in the therapy and treatment of such diseases, by creating accurate models of human disease for insightful and detailed studies.

5 Ways to Interpret the Myriad of Biological Data

To gain a wide perspective of the particular gene mutation consequences, we have to correlate the information obtained from experimental analysis with clinical data. Particularly taking into account the sequencing data from patient genomes. Such data become more and more available since nowadays whole exome and genome sequencing are often incorporated into personal health care. To achieve this goal, in the USA the Undiagnosed Disease Network (UDN) was established. The UDN brings together clinical and research data to learn more about genetic diseases, especially to help patients affected by unknown conditions. The data from patient genomes provides information regarding gene variants. The gene sequence can differ among populations because genes are mixed during sexual reproduction. Additionally, mutations are also a source of genetic variation. The interpretation of sequencing data, especially in the case of rare genetic variants, is difficult, because we do not know which individual gene or gene variant is responsible for disease. Particular gene variants can potentially lead to disease, and this possibility can be tested experimentally. Therefore, research institutes, such as the Model Organism Screening Center, use model organisms to check the pathogenicity of certain genetic variants identified in patients by the UDN.

Understanding the impact of gene variants would help us to increase the efficiency of diagnosis. To find a "good candidate gene for disease" the scientists have to combine and compare several sets of data, from different resources. This information can be gained from genetic data sets, but also molecular, morphological, and phenotypic data by the use of genetic animal models. The one of approaches called the MARRVEL (Model Organism Aggregated Resources for Rare Variant Exploration[2]) aims to combine public genetic resources to analyze and choose the gene variants, which most likely are the cause of certain human diseases. Such variants became a candidate for study in model organisms, for in-depth analysis of pathogenic mechanisms. To facilitate this process and gather all the data in one database, MARRVEL extracts data from clinical and scientific public databases and aligns the variants with model organism genetic sequences. Additional biological and experimental data can be gained from public databases regarding particular organisms. For example, FlyBase[3] is the most popular database describing fruit fly molecular data, and ZFIN[4] contains information concerning zebrafish.

The whole MARRVEL approach starts with a patient, who has a disease symptom with a potential genetic background. The results of Whole Exome Sequencing or Whole Genome Sequencing leads to a long list of candidate gene variants that can cause disease in this particular case. The data from different sources help to assess how likely a particular variant can be a reason for such a genetic disorder. Then the scientist checks if a particular variant has its homolog (counterpart) in the model organism. Sometimes, there is experimental data regarding this variant, which facilitates its assessment. Having such information, scientists are often able to prioritize the variants' pathogenicity. If there is no data, gene variants can be introduced into the model organism by genetic engineering. The animal genetic model is observed, tested and analyzed with the hope of finding similar phenotypic features, to those observed in the patient. This validation can show if the variant is pathogenic or not. Further, it can help to find appropriate treatment.

6 How Can Fruit Fly Help You?

To visualize the importance of this huge international scientific effort, we can use the true story as an illustration. Let's imagine a patient, a small girl, with severe disease symptoms, yet undiagnosed. Bristol was born as a healthy baby, but during the first three months, her parents noticed that her development was not as fast as it should be. Her body was floppy and flexible, she was not able to lift her head, and sometimes her left eye turned inwards. She was also delayed in further progress such as crawling, standing and walking. She cannot smile and seems to be not sensitive to pain. For example, she did not cry when she fell or get her vaccine shots. This girl was diagnosed with congenital hypotonia, a symptom that can be caused by different neurological or non-neurological conditions. To find a reason for Bristol hypotonia, for five years she underwent tests checking for cerebral palsy, Down syndrome, autism, muscular dystrophy and many others, all of which were negative. Finally, thanks to analysis within the Undiagnosed Disease Project, the doctors found two more children, a boy and a girl, with similar symptoms, who carried an identical point mutation in the Early B-Cell Factor 3 (EBF3)-coding gene. The *EBF3* gene is responsible for the regulation of various other genes' expression, therefore it is highly possible that its alterations can widely influence nervous and muscle systems. Therefore, the researchers decided to check the consequences of *EBF3* gene mutation in the fruit fly model. The experimental outcomes lead to the conclusion that *EBF3* is indeed responsible for disease symptoms observed in small patients. The additional 20 patients all around the world have been found to carry the mutation(s) in this gene, leading to the conclusion that it can be responsible for several cognitive or speech disorders, and hypotonia of unknown origin. The knowledge regarding the disease's origin allows for easier diagnosis in the future saving families from the "diagnosis odyssey": a long, uncertain and difficult journey through a variety of specialists and medical tests. Additionally, animal models can be used to test potentially therapeutic compounds leading to the discovery of efficient treatment (Dunlap, n.d.).

7 Animal Models Limitations

Animal models have great importance in the life sciences, however, these models are not devoid of disadvantages. There is no such animal model, which could mimic all features of human disease, simply because we are non-identical. The differences between species cause differences in phenotype, morphology, molecular or physiological pathways, etc. Therefore, it is important to choose the best animal to perform particular research to gain as informative outcome as possible. As mentioned before, the homology between human and animal genes is the key issue when you plan to model human genetic diseases. But other features are also important. The zebrafish are a very good model for drug screening, and phenotype-based drug discovery, pharmacology, and toxicology. Fishes are small, and produce a large number of eggs (one couple can produce 100 eggs, twice a week), which can be kept in small vessels. The transparent body at the early stages of development allows for easy, semi-automatic assessment of fish phenotype. Additionally, there is a growing number of available tools for phenotype assessment. However, the drugs or chemical compounds have to be water-soluble, and used in quite large quantities, since we have to immerse fish into the solution. Zebrafish embryos are protected by an envelope, called chorion, which keeps away many chemicals. Therefore, the experimental procedure requires additional steps, among others, special chemical treatment, to remove chorion. Moreover, the zebrafish development is external, which means that embryos do not have a placenta. Additionally, small animals are not good for monitoring biochemical parameters in blood, since taking samples from tiny organisms is very difficult or just not possible. Also, the drug doses tested in animals cannot be directly related to human physiology. Here we talk about zebrafish, which is only one example, but similar considerations have to be taken into account in every animal model and each experiment. Therefore, careful planning is crucial during the experiment designing and result interpretation (Dubińska-Magiera et al., 2021; Hartung, 2008; Migocka-Patrzałek & Elias, 2021; Robinson et al., 2019).

8 Ethical Issues

Experimentation on animals and ethical issues connected with such activity has been a topic of wide-world debate for decades. The necessity of conducting such experiments, which saves human lives, is undoubtful. However, the price is high because it causes animal distress. The discussion focusing on ethical issues leads scientists to establish rules, which allow them to avoid unnecessary experiments, and protect animals from suffering. In 1959 William Russell and Rex Burch described ethical guidelines, called the three R's (3Rs). The "R" letters relate to the terms Replacement, Reduction, and Refinement – three main, important practices, which should be applied during planning and conducting experiments involving animals (Tannenbaum & Bennett, 2015).

The first "R" says that, if it is possible the researcher should replace the experiment on animals with another technique. For example, one can use cell cultures (in vitro experiments) for the initial assessment of particular chemical toxicity. Another example is replacing animals with the so-called "artificial skin" (derived from in vitro cultures) in cosmetics testing. The alternative to using animals is the growing potential of bioinformatics analysis. Such in silico analysis allows for the virtual modelling of certain biological processes before testing them with the use of animals. For instance, one can use the prediction of protein-protein interaction to choose the best experimental conditions and parameters.

The second "R" obliged the investigator to reduce the number of animals used in the experiment to a minimum. This approach can be achieved by careful planning, also by using computational tools to calculate the minimal number of individuals in tested groups. Another example of following the "reduction" rule, is obtaining as much data from tested animals as possible. Let's imagine that we treat fish with a potentially therapeutic substance to observe overall phenotype. We can assess the changes using a variety of tests, for example, record a movie showing the swimming speed and pattern, make a photo of live fish to show its length and shape, and finally use the animal muscles for biochemical tests. In this way, we save at least two or three experimental groups. Additionally, we can use animal tissues, which were not used in experiments, for didactic purposes or other kinds of investigations and research.

The third "R" regards animal experiment refinement to avoid or ameliorate the animal suffering, distress and pain. This is an especially subtle issue because it is really hard to assess if any human intervention in animal natural life rhythms is harmless. Even tests taken as harmless, such as microscopic observations of live animal behaviour, introduce into their lives some additional stimuli such as different light, additional sounds or physical vibrations. All procedures, which could be painful, are carried out on appropriately anaesthetized animals (anaesthetic agents cause loss of feeling in all or part of the body, with or without loss of consciousness). But there are also several other ways to ameliorate the distress, that is. by maintaining stable, proper conditions in the breeding facility. It is also worth noting here, that the animals for research purposes are bred and maintained in proper, strictly controlled conditions. The animal facilities are obliged to keep rigid conditions such as, in the case of zebrafish, a certain quality of water and food, stable temperature, adequate aquatic systems, enough space and take into account many other important factors.

The 3Rs were stated to serve as a base for the development of future alternatives to the use of animals in research. Indeed, nowadays there are expanding range of options to replace experimentations on animals such as 3D cell cultures, computer modelling and bioinformatic tools (Andersen & Winter, 2019; Balls, 2020; Díaz et al., 2020; Petetta & Ciccocioppo, 2021; Robinson et al., 2019).

9 Conclusions

Simple organisms such as bacteria, yeasts, worms, and fruit flies are indispensable to gaining knowledge regarding human organisms and health. The model means something simpler than the original, which we used as an example. The fast development of alternative experimental systems will lead us to use in the future other than animal disease models, at least in some cases (Benam et al., 2015). Nevertheless, the use of animals as a model of the human body is of utmost importance in drug testing and medicine, simply because we, as mankind, are not skilled enough to imitate nature in a sufficient way. The simplest organism created by nature is far too complex to replace it fully with any artificial creation.

Abbreviations

ZNFS	zinc finger nucleases
TALENS	transcription activator-like effector nucleases
CRISPR	clustered regularly interspaced short palindromic repeats
Cas9	caspase 9

Notes

1 https://www.jax.org/
2 https://marrvel.org/; retrieved October 6, 2021
3 https://flybase.org/; accessed June 24, 2024
4 https://zfin.org/; accessed June 24, 2024

References

Andersen, M. L., & Winter, L. M. F. (2019). Animal models in biological and biomedical research – experimental and ethical concerns. *Anais Da Academia Brasileira de Ciencias*, *91*(suppl 1), e20170238.

Balls, M. (2020). It's time to reconsider the principles of humane experimental technique. *Alternatives to Laboratory Animals: ATLA*, *48*(1), 40–46.

Benam, K. H., Dauth, S., Hassell, B., Herland, A., Jain, A., Jang, K.-J., Karalis, K., Kim, H. J., MacQueen, L., Mahmoodian, R., Musah, S., Torisawa, Y.-S., van der Meer, A. D., Villenave, R., Yadid, M., Parker, K. K., & Ingber, D. E. (2015). Engineered in vitro disease models. *Annual Review of Pathology*, *10*, 195–262.

Burggren, W. W., & Warburton, S. (2007). Amphibians as animal models for laboratory research in physiology. *ILAR Journal/National Research Council, Institute of Laboratory Animal Resources, 48*(3), 260–269.

Chenouard, V., Remy, S., Tesson, L., Ménoret, S., Ouisse, L.-H., Cherifi, Y., & Anegon, I. (2021). Advances in genome editing and application to the generation of genetically modified rat models. *Frontiers in Genetics, 12*, 615491.

Cordier, A. C., & Haumont, S. M. (1980). Development of thymus, parathyroids, and ultimo-branchial bodies in NMRI and nude mice. *The American Journal of Anatomy, 157*(3), 227–263.

Díaz, L., Zambrano, E., Flores, M. E., Contreras, M., Crispín, J. C., Alemán, G., Bravo, C., Armenta, A., Valdés, V. J., Tovar, A., Gamba, G., Barrios-Payán, J., & Bobadilla, N. A. (2020). Ethical considerations in animal research: The principle of 3R's. *Revista de Investigacion Clinica; Organo Del Hospital de Enfermedades de La Nutricion, 73*(4), 199–209.

Dickinson, M. E., Flenniken, A. M., Ji, X., Teboul, L., Wong, M. D., White, J. K., Meehan, T. F., Weninger, W. J., Westerberg, H., Adissu, H., Baker, C. N., Bower, L., Brown, J. M., Caddle, L. B., Chiani, F., Clary, D., Cleak, J., Daly, M. J., Denegre, J. M., ... Murray, S. A. (2016). High-throughput discovery of novel developmental phenotypes. *Nature, 537*(7621), 508–514.

Dubińska-Magiera, M., Daczewska, M., Lewicka, A., Migocka-Patrzałek, M., Niedbalska-Tarnowska, J., & Jagla, K. (2016). Zebrafish: A model for the study of toxicants affecting muscle development and function. *International Journal of Molecular Sciences, 17*(11). https://doi.org/10.3390/ijms17111941

Dubińska-Magiera, M., Migocka-Patrzałek, M., & Cegłowska, A. (2022). Danio adventure. Developmental biology of the zebrafish in science popularisation. *Journal of Biological Education, 56*(3), 245–255.

Dubińska-Magiera, M., Migocka-Patrzałek, M., Lewandowski, D., Daczewska, M., & Jagla, K. (2021). Zebrafish as a model for the study of lipid-lowering drug-induced myopathies. *International Journal of Molecular Sciences, 22*(11).

Dunlap, E. (n.d.). *How fruit flies helped to find a diagnosis for my daughter.* Texas Children's Hospital. https://nri.texaschildrens.org/blog/how-fruit-flies-helped-find-diagnosis-my-daughter

Ericsson, A. C., Crim, M. J., & Franklin, C. L. (2013). A brief history of animal modeling. *Missouri Medicine, 110*(3), 201–205.

Gondo, Y., Fukumura, R., Murata, T., & Makino, S. (2009). Next-generation gene targeting in the mouse for functional genomics. *BMB Reports, 42*(6), 315–323.

Groza, T., Gomez, F. L., Mashhadi, H. H., Muñoz-Fuentes, V., Gunes, O., Wilson, R., Cacheiro, P., Frost, A., Keskivali-Bond, P., Vardal, B., McCoy, A., Cheng, T. K., Santos, L., Wells, S., Smedley, D., Mallon, A.-M., & Parkinson, H. (2023). The International

Mouse Phenotyping Consortium: comprehensive knockout phenotyping underpinning the study of human disease. *Nucleic Acids Research*, *51*(D1), D1038–D1045.

Hartung, T. (2008). Thoughts on limitations of animal models. *Parkinsonism & Related Disorders*, *14*(Suppl 2), S81–S83.

Hau, J. (2008). Animal models for human diseases. In P. M. Conn (Ed.), *Sourcebook of models for biomedical research* (pp. 3–8). Humana Press.

Kiełbwna, L., & Migocka-Patrzałek, M. (2017). Models of amphibian myogenesis – the case of Bombina variegata. *The International Journal of Developmental Biology*, *61*(1–2), 17–27.

Migocka-Patrzałek, M., & Elias, M. (2021). Muscle glycogen phosphorylase and its functional partners in health and disease. *Cells*, *10*(4). https://doi.org/10.3390/cells10040883

Muñoz-Fuentes, V., Cacheiro, P., Meehan, T. F., Aguilar-Pimentel, J. A., Brown, S. D. M., Flenniken, A. M., Flicek, P., Galli, A., Mashhadi, H. H., Hrabě de Angelis, M., Kim, J. K., Lloyd, K. C. K., McKerlie, C., Morgan, H., Murray, S. A., Nutter, L. M. J., Reilly, P. T., Seavitt, J. R., Seong, J. K., ... The IMPC Consortium. (2018). The International Mouse Phenotyping Consortium (IMPC): A functional catalogue of the mammalian genome that informs conservation. *Conservation Genetics*, *19*(4), 995–1005.

Onaciu, A., Munteanu, R., Munteanu, V. C., Gulei, D., Raduly, L., Feder, R.-I., Pirlog, R., Atanasov, A. G., Korban, S. S., Irimie, A., & Berindan-Neagoe, I. (2020). Spontaneous and induced animal models for cancer research. *Diagnostics (Basel, Switzerland)*, *10*(9). https://doi.org/10.3390/diagnostics10090660

Petetta, F., & Ciccocioppo, R. (2021). Public perception of laboratory animal testing: Historical, philosophical, and ethical view. *Addiction Biology*, *26*(6), e12991.

Plantié, E., Migocka-Patrzałek, M., Daczewska, M., & Jagla, K. (2015). Model organisms in the fight against muscular dystrophy: Lessons from drosophila and Zebrafish. *Molecules*, *20*(4), 6237–6253.

Robinson, N. B., Krieger, K., Khan, F. M., Huffman, W., Chang, M., Naik, A., Yongle, R., Hameed, I., Krieger, K., Girardi, L. N., & Gaudino, M. (2019). The current state of animal models in research: A review. *International Journal of Surgery*, *72*, 9–13.

Rosenthal, N., & Brown, S. (2007). The mouse ascending: perspectives for human-disease models. *Nature Cell Biology*, *9*(9), 993–999.

Szadvari, I., Krizanova, O., & Babula, P. (2016). Athymic nude mice as an experimental model for cancer treatment. *Physiological Research / Academia Scientiarum Bohemoslovaca*, *65*(Suppl 4), S441–S453.

Tannenbaum, J., & Bennett, B. T. (2015). Russell and Burch's 3Rs then and now: The need for clarity in definition and purpose. *Journal of the American Association for Laboratory Animal Science: JAALAS*, *54*(2), 120–132.

CHAPTER 15

The Beauty and Simplicity of the Complex Biological Chemistry of Metals

Magdalena Rowińska-Żyrek

Abstract

How to talk about complicated things in a simple way? To what extent can we simplify complex processes without simplifying them too much? I will try to bring us closer to answering these questions, referring to examples from my field of study. In both simple and complex ways, I will discuss the interaction of metal ions with a variety of biomolecules: proteins essential for the virulence of bacteria and fungi, peptides involved in neurodegeneration and the human ribozyme, an enzyme-like molecule consisting only of RNA. Briefly, I will try to convey the beauty and simplicity of the complex biological chemistry of metals.

Keywords

bioinorganic chemistry – thermodynamics – nickel chaperones – zincophores – CPEB3 ribozyme

•••

Most complex discoveries start with simple questions.

∴

I am not familiar with statistics, but the statement is certainly true for several big scientific discoveries, which started with the simplest of ideas and observations (gravity works, light travels pretty quickly, the level of water rises when we enter the bathtub, etc.). The explanation of these phenomena is usually more complex than the question itself, but without the easy question, without

 | DOI:10.1163/9789004714540_015

the driving force of human curiosity, there would have been no answer, no science and no progress.

"Making the simple complicated is commonplace; making the complicated simple, awesomely simple, that's creativity" (in: Nixon, 2020, p. 4). The author of these words, Charles Mingus, a legendary jazz player, just flattered a lot of my fellow scientists, suggesting that they are one of the most creative creatures on earth. I would not be that generous in my judgements, but I do have to agree that there is some truth in the statement.

Let us think of a deduction process composed of four stages: (1) a simple question; (2) a complicated analysis; (3) a simplification that allows understanding the problem and adjusting an experimental approach; (4) the simple answer.

Now let me make what I just wrote simpler, by showing some examples from my field of studies. In bioinorganic chemistry, the general, simple question is: why do living creatures need metal ions? And what happens, if something 'goes wrong' in their availability or distribution?

Sounds easy, right? This easy question led to an avalanche of discoveries, allowing us to understand insights into the molecular workings of metalloenzymes, anticancer drugs, the human brain and pathogenic bacteria. And it all started with simple questions.

Let me give you an example:

1 The Simple Question

Let's be more precise in the question we want to ask, and narrow the general "why do living creatures need metal ions?" to "why does the stomach ulcer-causing bacteria need nickel ions?". This relatively simple question is a consequence of scientific curiosity, inspired by the discovery of *Helicobacter pylori* (*H. pylori*). Now, thirty years after Barry Marshall drank a suspension of the bacteria to prove that it is the cause of stomach ulcers, it is obvious that is the Gram-negative *H. pylori*, which colonizes the gastric mucosa in humans, causing acute and chronic gastritis, peptic ulcer disease, gastric carcinoma, and gastric lymphoma (Kusters et al. 2006). It is estimated that about 40–50% of the adult population in developed countries, and up to 90% of adults in developing countries can be infected with *H. pylori* and new, non-antibiotic treatments are still being looked for.

Coming back to the topic: why does *H. pylori* need Ni^{2+} ions? What happens when it doesn't get any, and how can we make it happen?

2 The Complicated Analysis

The answer starts in a relatively simple way – *H. pylori* produces an assortment of factors in order to adapt to the extremely acidic environment of the stomach. We have to keep in mind, that although the bacterium successfully inhabits the extremely low pH of the stomach, it multiplies in an environmental pH from 6.0 to 8.0 and cannot survive at pH below 4.0 (Morgan et al., 1987).

To survive in those unwelcoming living conditions, *H. pylori* depends on two nickel-containing enzymes, urease and hydrogenase. The dinuclear Ni^{2+}-containing urease accounts for about 10% of the soluble cellular proteins (Bauerfeind et al., 1997), catalyzes the hydrolysis of urea into carbon dioxide and ammonia and therefore neutralizes the low gastric pH around the bacteria (Scott et al., 2002). The activity of urease critically depends on the availability of nickel – a functional urease complex requires 24 Ni^{2+} ions (Hawtin et al., 1991). Another factor is a membrane bound [NiFe] hydrogenase, which permits respiratory based energy production for the bacteria in the mucosa (Olson & Maier, 2002). Most of the bacterium's metal metabolism is centered upon the expression and maturation of those two Ni^{2+}- dependent enzymes (which is not a surprise, because the task is quite challenging – the concentration of Ni^{2+} in human plasma is as low as 0.44 nM) and a set of at least 22 accessory proteins involved in the homeostasis of nickel is needed for their activation (Mehta et al., 2003). Understanding this mechanism is a crucial basis for developing new, highly specific treatments that would aim at the "weak point" of this machinery. Since the number of people resistant to standard therapy (two antibiotics and a proton pump inhibitor) is rapidly increasing (over 20% do not respond to the therapy at all), new solutions in the treatment of ulcers are still being looked for. One of such therapies is a group of drugs based on bismuth (III) salts. Currently, there are three different salts available on the market – bismuth subsalicylate (in Pepto-Bismol), colloidal bismuth subcitrate (in De-Nol) and ranitidine bismuth citrate are commonly used for the treatment of peptic ulcers, caused by the presence of *H. pylori*. How do they work? It is tempting to think that Bi^{3+} is able to displace Ni^{2+} from its binding sites. Which proteins look as if they could bind both metals? How can we find out how bismuth ions work, without getting lost in the complicated network of the large and numerous nickel-binding proteins (Figure 15.1)?

What should we start with, trying to answer our 'simple' (now it sounds ironic…) question? A good starting point would be a protein that has potential good metal binding sites in unstructured regions.

HspA, Hpn and Hpn-like are good examples of such.

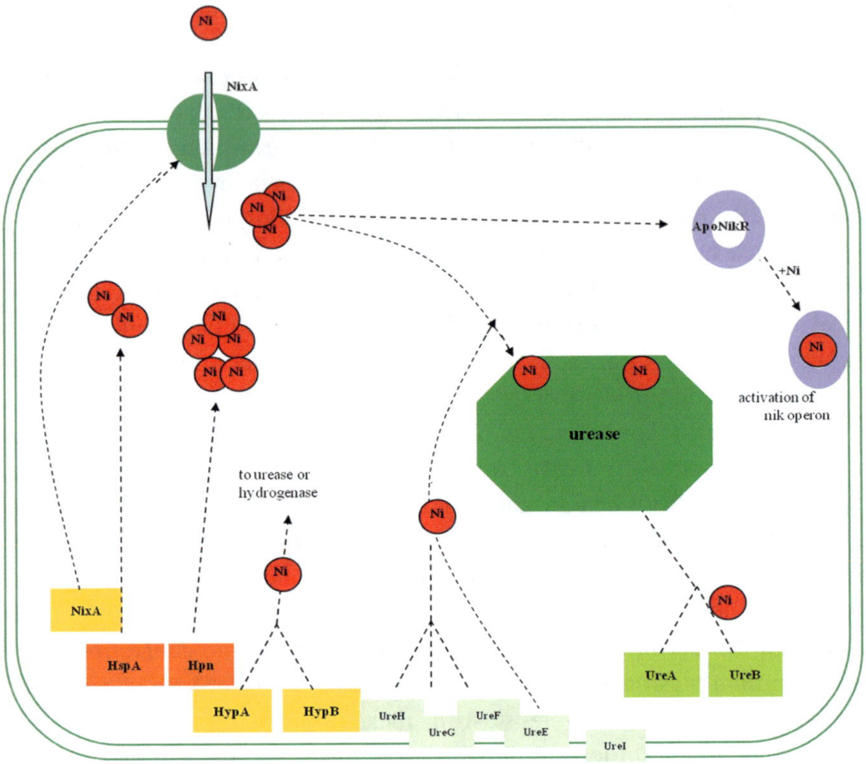

FIGURE 15.1 The complex system of nickel homeostasis in *H. pylori*

Heat Shock Protein A (HspA) is a GroES homolog involved in the folding of polypeptide chains into oligomeric complexes. In the case of *H. pylori,* HspA has a second function, being involved in the homeostasis of nickel. It delivers nickel to appropriate proteins in a Ni^{2+}-deficient environment and takes part in the detoxification through the sequestration of excess nickel. Under normal conditions, HspA delivers Ni^{2+} to proteins that have a higher affinity towards this metal than HspA itself. In *H. pylori*, the protein consists of 118 amino acids divided into two domains: an N-terminal domain (domain A, residues 1–90), which is homologous with other GroES bacterial proteins, and an unstructured histidine and cysteine-rich C-terminal domain (domain B, residues 91–118), which other GroES-like proteins do not have (Suerbaum et al., 1994) (Figure 15.2-A). This is a good indication that this is the domain we want to take a closer look at in our 'simplified' approach.

Two other proteins that could be 'suspected' of being excellent nickel binders are the extremely His-rich Hpn and Hpn-like, two of *Helicobacter pylori's* cytoplasmic proteins involved in the homeostasis of nickel. Almost half of Hpn's amino acid sequence (28 out of 60 residues) are histidines (Figure 15.3). Hpn-like, highly similar in its primary sequence to Hpn, contains some additional polyglutamine repeats (25% of its sequence are histidines, while 42% of the amino acid residues are glutamines). Both proteins are not only urease and hydrogenase nickel accessory proteins but also important factors in nickel (and other metals) storage and detoxification (Li & Zamble, 2009). Hpn is abundant in the cytoplasm of *H. pylori* and accounts for approximately 2% of all proteins synthesized by the bacterium (reviewed in: Mulrooney & Hausinger, 2003; Kaluarachchi et al., 2010).

Now, having those proteins in mind, let us again ask: what happens, when something 'goes wrong' with nickel distribution in *H. pylori*? To answer the simple question, it will be necessary to make several simplifications of the complex system.

3 The Simplification

In this case, the simplification is based on working with model systems – peptides, which are representative unstructured fragments of the studied nickel binding proteins, 'suspected' of being the metal binding site. Simple: the idea is to work with the metal binding site, not with the full protein, and to precisely determine the affinity binding constants to each of the potential metal binding sites or to investigate peptide recognition. Or, even simpler: we chop up the protein into smaller pieces and we check where the metal binds most strongly or we check if the two regions interact with each other.

This simplification has several limitations that we have to be aware of. It would not be particularly useful if the studied protein region had a pre-defined structure, or was a membrane protein (I will come back to this point later, now the plan is to make things simple; the verification process will follow).

Let us first focus on the C-terminal part of HspA. Its eight His and four Cys residues among 27 amino acids (GSCCHTGNHDHKHAKEHEACCHDHKKH) make it a tempting nickel binder. Keeping in mind, that the N-terminal domain has the function of a heat shock protein, and is not involved in nickel homeostasis, and that the C-terminal domain (absent in other species) is unstructured and very probable to chelate metal ions, it is reasonable to focus on domain B only. The shorter the studied region, the more precise thermodynamic data we are able to obtain about its metal complex (not even to mention the lower costs and easier handling of the sample …).

Indeed, Ni^{2+} ions make a thermodynamically strong square planar complex with two sulfurs of neighboring cysteine residues and to the amide between

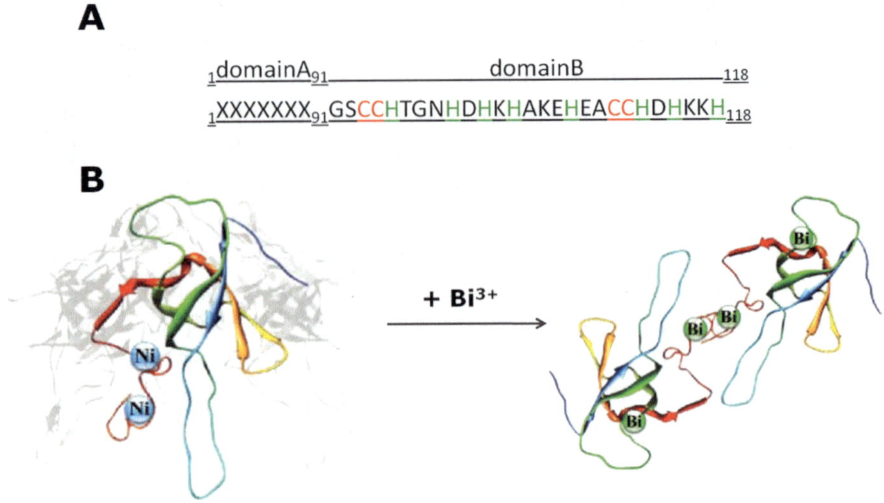

FIGURE 15.2 A. A scheme of Heat Shock Protein A from *H. pylori*
B. A suggested scheme of the structural rearrangement of HspA upon Bi^{3+} addition; only the front subunit is highlighted (Cun et al., 2008)

them. Looking at the interaction of bismuth with the same model system, at least one of the ways how bismuth might inhibit nickel-dependent proteins of this bacterium becomes obvious – thermodynamically, bismuth complexes are over ten orders of magnitude more stable than nickel ones (Rowińska-Żyrek et al., 2009). Moreover, Bi^{3+} is able to displace Ni^{2+} from its Cys-rich binding sites (Rowińska-Żyrek et al., 2010), disrupting the homeostasis of nickel and/ or causing nickel deficiency during post-translational modifications of urease and hydrogenase.

The finding was confirmed by the biological study, which showed that Bi^{3+} induces the changes in quaternary structure of the protein, making it refold from a native heptamer to a dimer (Cun et al., 2008). The binding is irreversible at physiological pH, which clearly indicates that bismuth may inhibit the biological functions of HspA (Cun et al., 2008).

How about the abnormally histidine-rich protein, Hpn? It mostly exists as an unstructured multimer in solution, with each 7 kDa monomer binding 5 nickel ions at pH 7.4 (Ge et al., 2006). Bacteria lacking Hpn, cultured in vitro have an increased response to therapeutic forms of bismuth (Mobley et al., 1999). How does this happen, what are the details of Ni^{2+} and Bi^{3+} coordination to this protein?

A striking feature of Hpn, apart from the impressive amount of histidines, is a variety of available binding sites for metal ions. Nickel, native to the protein, is able to bind to the ACTUN motif in an albumin-like binding (MAH), to the His-rich motif (HHH) and to the cysteine thiols (CC). Since Hpn does not have a defined structure and all of the mentioned sites are available for binding, this particular protein is an excellent example based on which we can debate on the thermodynamic preferences of nickel binding. What is the outcome of the debate? The N-terminal part of Hpn binds Ni^{2+} ions in the same way as native human albumin does, forming nickel and copper complexes involving a $\{NH_2, 2N^-, N_{im}\}$ donor set. Two other fragments with a -CC- motif (Ac-CCSTSDSHHQ-NH_2 and Ac-EEGCCHGHHE-NH_2) derived from the C-terminal part of Hpn, bind nickel, forming very stable planar complexes with both peptides, due to the $\{2S^-,N^-\}$ binding mode (Rowińska-Żyrek et al., 2013).

The 'simplified' approach allowed an understanding of the behaviour of a polyhistidyl fragment of Hpn (Ac-THHHHYHGG-NH_2) in a complex with Ni^{2+}. The coordination properties of this so-called wild-type fragment were compared with those of its six analogues, in which consecutive residues (His or Tyr) were replaced by Ala (Ala-substitution or Ala-scan approach) resulting in Ac-TAHHHYHGG-NH_2, Ac-THAHHYHGG-NH_2, Ac-THHAHYHGG-NH_2, Ac-THHHAYHGG-NH_2, Ac-THHHHAHGG-NH_2 and Ac-THHHHYAGG-NH_2 peptides, respectively. This approach allowed to conclude that the fourth His residue is critical for the binding of Ni^{2+} and that the effectiveness of binding varies even if the substituted amino acid does not directly bind. Moreover, we showed that the metals cannot bind to four consecutive histidines (Witkowska et al., 2012).

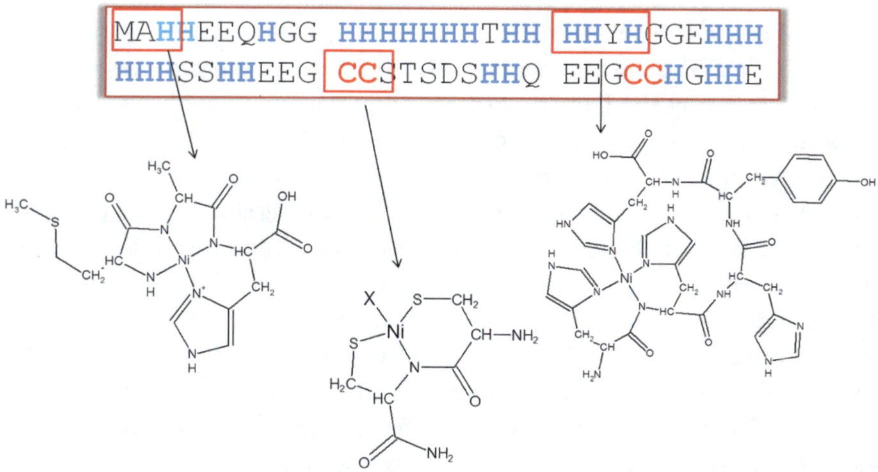

FIGURE 15.3 Binding modes of nickel to the Hpn sequences discussed above – HHYH, MAH and CC

One of the most useful features of the "peptide-based approach" is the precision with which we can compare the thermodynamic differences of the different binding sites. Among the three different parts of Hpn mentioned above, it is the C-terminal part of the protein with the "CC" sequence which binds nickel most effectively. The His-rich fragment binds nickel more efficiently at lower pH than the N-terminal part does; at physiological pH (7.4), we observe the contrary.

Another fun fact about Hpn and Hpn-like: the more glutamine residues they have, the stronger they coordinate metal ions. The 'simplified' approach was used to study the interactions of Ni^{2+} with several N-terminal domains of Hpn and Hpn-like proteins from *H. pylori*, each with a different number of glutamine residues, but with the same albumin-like metal binding mode. Experimental results, in very good agreement with theoretical findings, lead to the not-at-all obvious conclusion that the stability of metal complexes distinctly increases with the number of glutamine residues present in the peptide, although glutamine side chains do not directly take part in coordination. Most probably forms a network of hydrogen bonds which protects the metal binding site from water molecules. This peculiar finding allows one to look at polyglutamine sequences, not only the ones present in some bacterial chaperons but also those involved in several neurodegenerative diseases, from a new perspective (Chiera et al., 2013).

4 The Simple Answer

Helicobacter pylori depends on urease and hydrogenase to survive in the acidic environment of the human stomach. Both of the two enzymes need nickel ions, which is why the bacteria express numerous proteins that ensure the proper distribution and storage of Ni^{2+}. Moving in the world of *H. pylori*'s protein interactions with metals, using model systems – unstructured protein fragments, which are 'suspected' of being metal-binding sites, we can point out the regions with the highest affinity towards nickel and prove that the component of commonly used antiulcer drug, bismuth, may inhibit nickel dependent proteins of this bacterium – thermodynamically, bismuth complexes are over ten orders of magnitude more stable than nickel ones and Bi^{3+} is able to displace Ni^{2+} from its Cys-rich binding sites (Rowińska-Żyrek et al., 2010), disrupting the homeostasis of nickel and causing nickel deficiency during post-translational modifications of urease and hydrogenase.

'*Making* the complicated simple, *awesomely simple*'– *a series of cases*

Clearly, *H. pylori* related studies are not the only ones to which the "simple question-complicated analysis-simplification-the simple answer" deduction scheme can be applied. Even more than that – the system doesn't even have to be a protein. Examples? With pleasure – a ribozyme example. The "divide and conquer" policy told us a lot about the correlation between the structure and function of one of the only three known human small catalytic RNAs, the CPEB3 ribozyme (Figure 15.4A).

Until recently, ribozymes ('enzymes' composed solely of RNA) were found only in lower organisms. The discovery of the human CPEB3 ribozyme, related to the Hepatitis Delta Virus (HDV, Figure 15.4B) one, was an enormous breakthrough.

This ribozyme is a highly conserved, mammalian, self-cleaving, non-coding RNA located in the second intron of the *cpeb3* gene (Salehi-Ashtiani et al., 2006). This gene encodes a cytoplasmic polyadenylation element binding protein that promotes the elongation of the polyadenine tail of messenger RNA and mediates germ cell development, synaptic plasticity, learning and memory; it has also been suggested to adapt prion-like conformations. The CPEB3 protein is rather well studied, yet surprisingly little is known about the CPEB3 ribozyme itself. Most of the available information is based on comparative studies with the structurally and biologically related HDV ribozyme. Despite having divergence in base sequences, both the viral HDV and the human CPEB3 were predicted to form the same base pairing interactions, which result in a pseudoknot structure (Webb et al., 2009). How to verify the formation of the predicted nested double pseudoknot in solution and how to get an understanding of the structure-function and metal ion binding-function relationships of this 67 nucleobase-long ribozyme? The NMR structure of the whole construct is highly desirable (and highly complicated to obtain), as well as NMR-monitored Mg^{2+} titrations, to characterize specific metal binding sites and structural changes within the ribozyme upon addition of the metal ions.

NMR proton resonance assignment of the CPEB3 ribozyme was a huge challenge because of the severe spectral overlap in the [^{1}H,^{1}H]-NOESY spectra, which is mainly due to the large size of the ribozyme. All assignments were made in spectra of partially deuterated CPEB3 RNA, in which line widths were drastically improved compared to the spectra of natural abundance samples. The assignment was supported by using different labelling schemes for the CPEB3 RNA, which helped to select for [^{1}H,^{1}H]-NOESY or [^{1}H,^{15}N]-HSQC correlations belonging to certain nucleotides. Four samples with only one NMR-visible (natural abundance) nucleotide (with the remaining three being fully deuterated ones) were also analyzed (Rowińska-Żyrek et al., 2014a).

Further NMR analysis showed that the global fold of the CPEB3 ribozyme, encompassing 80–90% of the predicted base pairs, is formed in monovalent ions alone. Low millimolar concentrations of Mg^{2+} promote a more compact fold and lead to the formation of additional structure in the core of the ribozymes, which contains the inner small pseudoknot and the active site. Several Mg^{2+} binding sites, which are important for the functional fold, appear to be located in corresponding locations in the HDV and CPEB3 ribozyme, demonstrating the particular relevance of Mg^{2+} for the nested double pseudoknot structure.

Still, no clue about structure? Keep it simple; apply the 'divide and conquer' strategy – use three small model constructs for the P1 and P2 helices and for the P4 hairpin (Figure 15.4C). This simple idea was an absolute hit – results obtained for the three separate regions allowed solving their structures and facilitated the assignment of the full-length CPEB3 ribozyme spectra (Rowińska-Żyrek et al., 2014b).

Another example of an approach that makes the complicated simple? Understanding the interactions of Zn^{2+} with zincophores and zinc transporters from *Candida albicans* (*C. albicans*), the most common cause of fungal infections in humans.

C. albicans, although usually a commensal fungus, it is the most common cause of candidiasis – a condition that encompasses infections that range from

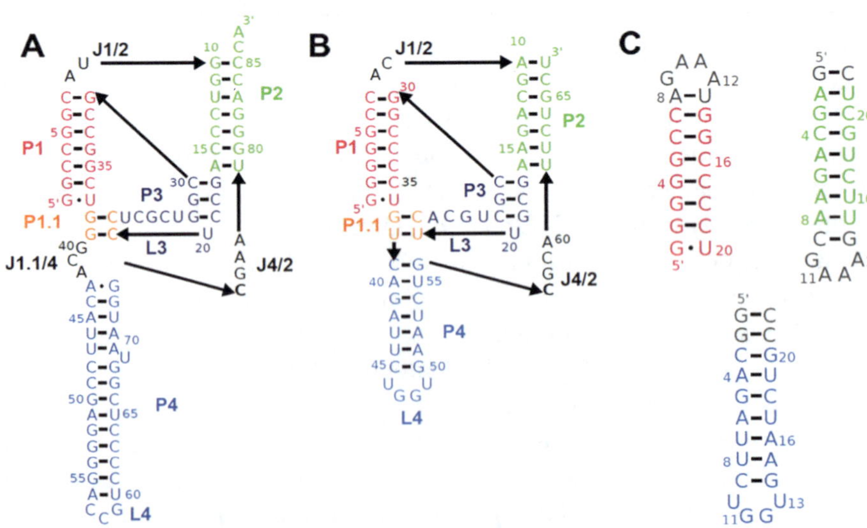

FIGURE 15.4 Secondary structures of (A) the genomic HDV ribozyme (B) the CPEB3 ribozyme (C) the small model constructs P1, P2 and P4 used for resonance assignment; the nucleotides added to the natural sequences are shown in grey

superficial, chronic to systemic and potentially life-threatening candidemia. The frequency of drug-resistant invasive mycoses has increased significantly over the past three decades, and is becoming a serious medical problem not only in immunosuppressed patients. As a result, new antifungal drugs and ways to specifically deliver them to the fungal cell are being looked for. One of the biggest obstacles in finding such comes from the fact that fungi share essential metabolic pathways with humans. One significant difference in the metabolism of those two cells that can be challenged when looking for possible selective therapeutics is the uptake of zinc (crucial for fungal survival and virulence).

Candida albicans relies on a mechanism of zinc uptake. It is based on a secreted protein, which specifically binds Zn^{2+}, Pra1, the so-called 'zincophore', a small, 299 amino acid secreted zinc-binding protein, which can sequester this metal from the environment and re-associate with the fungus via a co-expressed, genetically-linked membrane transporter, Zrt1. Citiulo et al., (2012) elucidated the mechanism of *C. albicans* zinc acquisition from host cells: (1) after the host cell invasion Pra1 (pH-regulated antigen 1, the previously mentioned zincophore), is expressed due to the alkaline pH and low amount of soluble zinc of the intracellular environment (Sentandreu et al., 1998; Outten & O'Halloran, 2001); (2) the protein is secreted and released from the fungal cell surface, predominantly in the hyphal form; it is required for hyphal extension and causes endothelial damage of the host (De Bernardis et al., 1998); (3) it binds host cellular zinc (either free cytosolic or bound to host protein); and (4) returns to the fungal cell via physical interaction with Zrt1, a membrane transporter, to deliver the bound metal ion (Figure 15.5) (Citiulo et al., 2012).

How to understand the bioinorganic chemistry of this process, to point out the Zn^{2+} binding sites in both Pra1 and Zrt1 and understand the thermodynamics of the Pra1-Zn^{2+}-Zrt1 interaction?

As you might already be guessing, our approach involves working on both full-length proteins and model systems (unstructured parts of proteins) in order to identify those regions in Pra1, to which bind zinc with the highest affinity and those which are recognised by Zrt1. Several unstructured regions of Pra1 and Zrt1 can be used to perform a structural and thermodynamical analysis of their complexes with Zn^{2+}, being a stepping stone towards finding new, fungus-specific treatments based on parts of zincophores coupled with an imidazole- or triazole-based antifungal drugs.

Verification of the simplification procedure? The best one would be performing growth promotion studies, indicating whether these Pra1 fragments are able to transfer zinc into fungi, and therefore act like zincophores.

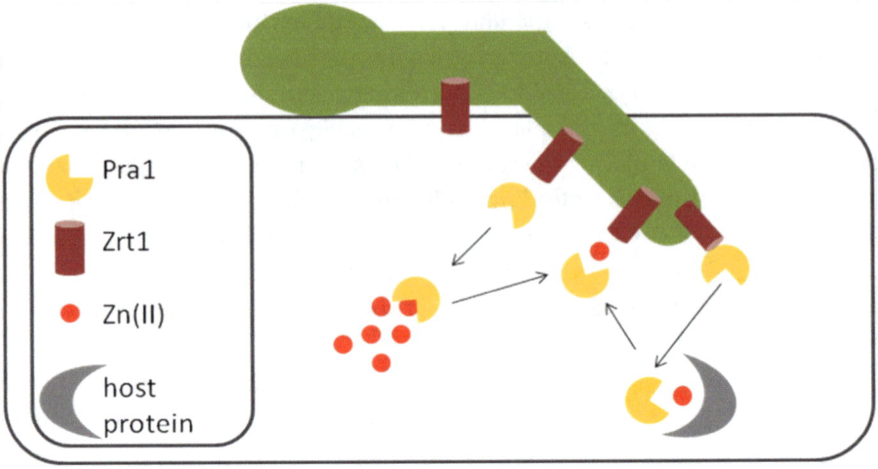

FIGURE 15.5 Schematic model of *C. albicans* zinc scavenging from host cells. After invasion of the host cell, Pra1 is expressed and secreted. It binds zinc, either in the form of free Zn^{2+} (extremely sparse in the cellular pool) or from zinc-binding proteins of the host. Reassociation with *C. albicans* cell surface and Zn^{2+} transport into the cell occurs via a Pra1-Zrt1 interaction

'Peptide-based', 'divide and conquer', however, we call the approach, if we are big fans of it or not, what is clear are two things: first, it was the key to several major discoveries – it would take time and severely abuse the dimensions of this book to describe all in brief (e.g., this simple approach allowed to elucidate the impact of metal ions on the molecular basis of Alzheimer's, Parkinson's or prion disease. Cu^{2+} and Fe^{3+} usually bind to specific sites in beta-amyloid, a-synuclein and prion proteins, resulting in misfolding and oxidative stress, which is the trigger for the diseases).

And second, it is of major importance to be aware of the obvious limitations of the method. A pre-defined structure of the biomolecule or a tertiary interaction with another group is a serious limitation.

Having started the chapter with a quotation, I will also finish it with one. I think one attributed to Albert Einstein fits: "Make everything as simple as possible, but not simpler" (Robinson, 2018).

References

Bauerfeind, P., Garner, R., Dunn, B. E., & Mobley, H. L. (1997). Synthesis and activity of *Helicobacter pylori* urease and catalase at low pH. *Gut, 40*(2), 25–30.

Chiera, N. M., Rowińska-Żyrek, M., Wieczorek, R., Guerrini, R., Witkowska, D., Remelli, M., & Kozłowski, H. (2013). Unexpected impact of the number of glutamine residues on metal complex stability. *Metallomics, 5*, 214–221.

Citiulo, F., Jacobsen, I. D., Miramion, P., Schild, L., Brunke, S., Zipfel, P., Brock, M., Hube, B., & Wilson, D. (2012). Candida albicans scavenges host zinc via Pra1 during endothelial invasion. *PLoS Pathog,* 8.

Cun, S., Li, H., Ge, R., Lin, M. C. M., & Sun, H. (2008). A histidine-rich and cysteine-rich metal-binding domain at the C terminus of heat shock protein A from Helicobacter pylori: implication for nickel homeostasis and bismuth susceptibility. *Journal of Biological Chemistry, 28*, 15142–15151.

De Bernardis, F., Muhlschlegel, F. A., Cassone, A., & Fonzi, W. A. (1998). The pH of the Host Niche controls gene expression in and virulence of *Candida albicans*. *Infection and Immunity, 66*, 3317–3325.

Ge, R., Zhang, Y., Sun, X., Watt, R. M., He, Q.-Y., Huang, J.-D, Wilcox, D. E., & Sun, H. (2006). Thermodynamic and kinetic aspects of metal binding to the histidine-rich protein. *Hpn. Journal of the American Chemical Society, 128*, 11330–11331.

Hawtin, P., Delves, H., & Newell, D. (1991). The demonstration of nickel in the urease of Helicobacter pylori by atomic absorption spectroscopy. *FEMS Microbiology Letters, 77*, 51–54.

Kaluarachchi, H., Chan Chung, K. C., & Zamble, D. B. (2010). Microbial nickel proteins. *Natural Product Reports, 27*, 681–694.

Kusters, J. G., van Vliet, A. H. M., & Kuipers, E. J. (2006). Pathogenesis of *Helicobacter pylori* infection. *Clinical Microbiology Reviews, 19*, 449–490.

Li, Y., & Zamble, D. B. (2009). Nickel homeostasis and nickel regulation: An overview. *Chemical Reviews, 109*, 4617–4643.

Mehta, N., Olson, J. W., & Maier, R. J. (2003). Characterization of *Helicobacter pylori* nickel metabolism accessory proteins needed for maturation of both urease and hydrogenase. *Journal of Bacteriology, 185*, 726–734.

Mobley, H. L., Garner, R. M., Chippendale, G. R., Gilbert, J. V., Kane, A. V.,& Plaut, A. G. (1999). Role of Hpn and NixA of *Helicobacter pylori* in susceptibility and resistance to bismuth and other metal ions. *Helicobacter, 4*, 162–169.

Morgan, D. R., Freedman, R., Depew, C. E., & Kraft, W. G. (1987). Growth of *Campylobacter pylori* in liquid media. *Journal of Clinical Microbiology, 25*, 2123–2125.

Mulrooney, S. B., & Hausinger, R. P. (2003). Nickel uptake and utilization by microorganisms. *FEMS Microbiology Reviews, 27*, 239–261.

Nixon, N. (2020). *The creativity leap: Unleash curiosity, improvisation, and intuition at work*. Berrett-Koehler Publishers.

Olson, J. W., & Maier, R. J. (2002). Molecular hydrogen as an energy source for *Helicobacter pylori*. *Science, 298*, 1788–1790.

Outten, C. E., & O'Halloran, T. V. (2001). Femtomolar sensitivity of metalloregulatory proteins controlling zinc homeostasis. *Science*, *292*, 2488–2492.

Robinson, A. (2018). Did Einstein really say that?. *Nature*, *557*(7703), 30–31.

Rowińska-Żyrek, M., Skilandat, M., & Sigel, R. K. O. (2014). Solution structure and metal ion binding sites of the human CPEB3 ribozyme's P4 domain. *Journal of Biological Inorganic Chemistry*, *19*(6), 903–912.

Rowińska-Żyrek, M., Skilandat, M., & Sigel, R. K. O. (2014). Secondary structure confirmation and localization of Mg^{2+} ions in the mammalian CPEB3 ribozyme, *RNA*.

Rowińska-Żyrek, M., Valensin, D., Szyrwiel, L, Grzonka, Z., & Kozłowski, H. (2009). Specific interactions of Bi(III) with the Cys-Xaa-Cys unit of a peptide sequence. *Dalton Transactions*, 9131–9140.

Rowińska-Żyrek, M., Witkowska, D., Potocki, S., Remelli, M., & Kozłowski, H. (2013). His-rich sequences is plagiarism from nature a good idea? *New Journal of Chemistry*, *37*, 58–70.

Rowińska-Żyrek, M., Witkowska, D., Valensin, D., Kamysz, W., & Kozłowski, H. (2010). The C terminus of HspA–A potential target for native Ni(II) and Bi(III) anti-ulcer drugs. *Dalton Trans*, *39*, 5814–5826.

Salehi-Ashtiani, K., Luptak, A., Litovchick, A., & Szostak, J. W. (2006). A genomewide search for ribozymes reveals an HDV-like sequence in the human CPEB3 gene. *Science*, *313*, 1788–1792.

Scott, D. R., Marcus, E. A., Weeks, D. L., & Sachs, G. (2002). Mechanisms of acid resistance due to the urease system of *Helicobacter pylori*. *Gastroenterology*, *123*, 187–195.

Sentandreu, M., Elorza, M. V., Sentandreu, R., & Fonzi, W. A. (1998). Cloning and characterization of PRA1, a gene encoding a novel pH-regulated antigen of *Candida albicans*. *Journal of Bacteriology*, *180*, 282–289.

Suerbaum, S., Thiberge, J. M., Kansau, I., Ferrero, R. L., & Labigne, A. (1994). *Helicobacter pylori* hspA-hspB heat-shock gene cluster: nucleotide sequence, expression, putative function and immunogenicity. *Molecular Microbiology*, *14*, 959–974.

Webb, C.-H. T., Riccitelli, N. J., Ruminski, D. J., & Luptak, A. (2009). Widespread occurrence of self-cleaving ribozymes. *Science*, *326*.

Witkowska, D., Politano, R., Rowińska-Żyrek, M., Guerrini, R., Remelli, M., & Kozłowski, H. (2012). The coordination of Ni(II) and Cu(II) ions to the polyhistidyl motif of Hpn protein: Is it as strong as we think? *Chemistry – A Eur Journal*, *18*, 11088–11099.

CHAPTER 16

On Simplificatory and Diversificatory Aspects of the Presentation of an Object

Bartłomiej Skowron

Abstract

In this article, I present and defend the view that our cognition of complex objects is on the one hand incomplete and unavoidably simplificatory, and on the other diversificatory, in the sense of conferring additional internal diversity on them by introducing features they do not otherwise have. To substantiate this viewpoint, I make use of the ontological analysis of both the object and its cognition carried out by the founding father of the Lviv-Warsaw School of philosophy, Kazimierz Twardowski, in his work *On the content and the object of presentations*. I also model Twardowski's findings topologically, using Fréchet's theory of types of dimension.

Keywords

Kazimierz Twardowski – object of presentation – content of presentation – material and formal parts – theory of an object – Maurice Fréchet's theory of types of dimension

1 Introduction

There are, we might be tempted to suppose, three protagonists: a subject, an object, and a relation of cognition that connects them. The subject gets to know the object, and the object becomes known by the subject. Getting to know the object is a complex process.[1] A first and naive approximation of that process might be the view that holds that within the subject certain presentations occur,[2] these being some form of mental duplication amounting to something like a photographic copy of the object that the subject seeks to know. The more exact the photographic duplication, the more adequate the cognition of the object.

 | DOI:10.1163/9789004714540_016

That viewpoint is false, for at least two reasons. Firstly, cognition cannot be confined to instances of photographic mirroring, as in that case how would we get to know such abstract objects as Hilbert spaces? What would the copies of these objects be?

Secondly, the totality of what counts as getting to know an object cannot be equated solely with the adequacy of that object's presentation. Presentations furnish the basic elements of cognition: they are the beginning of the cognition process, and as such do not themselves exhaust the entire cognition process. Amongst the components omitted by such an account there is, at the very least, the final result of cognition: i.e., the making of a judgment.

It is, in a certain sense, judgment that is the decisive factor (Ingarden, 1972), with the ineluctable role of the subject being to validate cognition. This is because the judgment we make about an object we seek to know constitutes the crowning moment of that object's cognition. Hence, the correct conclusion in this respect is that cognition does not create a photographic copy of the object of knowledge. But what, then, does it amount to?

Let us use the following analysis as a first approximation of the object's presentation, construed as the basis of cognition (Twardowski, 1977). The presentation of an object consists of three strongly interlinked components: the process (act) of presentation, the content of presentation and the object of presentation. The object of presentation is presented *through* the content of the presentation in the *act* of presentation. Our viewing the object of presentation through the content of the presentation is something to be analyzed via the decomposition of both object and content into their respective parts. The relationship between these parts can then be analyzed.

The relation obtaining naturally between some part of the content of presentation and some part of the object of presentation will be a certain correspondence. Should this be absent, the content of the presentation will not target the object, and the object of the presentation will not be adequate to the content of the presentation. We might expect to arrive at an understanding of that relation through observation and analysis of the results of their division into their respective parts: i.e. by differentiating the parts of the object of presentation and of the content of presentation. The key issues we aim to reflect on are the following: How do the parts of a presentation correspond with those of the object of presentation? How many parts of the content of presentation correspond with those of the object of presentation, and vice versa? In what way do the parts of the object of presentation correspond with those of the content of presentation? How many parts of the object of presentation correspond with the parts of the content of presentation? Our analysis will be carried out in the context of this sort of understanding of what correspondence amounts to.

The structure of the present article is therefore as follows: picking up from where Twardowski (1977) left off, I aim to render in more precise terms the concept of an object, to differentiate its parts, and to spell out in detail the relation of correspondence between the object and content of presentation. Next, I formulate and seek to justify my main thesis, to the effect that the presentation of complex objects is always a simplifying one, in the sense of one that omits some parts of the object, but – what is more – at the same time also an enriching one, in that it introduces into the object of presentation some parts which the latter does not itself have. The validation of this thesis proceeds via not one but two paths: through analysis of cognitive experience, and formal analysis of an object and content understood abstractly. The experience-based analysis is mainly based on the contribution of Twardowski himself, while the formal analysis aims to enhance our understanding the complexity of cognition through topological modeling, and in particular through the notion of *types of dimension* as introduced by Fréchet.

2 The Object and Content of a Presentation

Everything that is *something* is an object – that is, anything that can be presented. The totality of objects is coextensive with all that is *something*. In this wide sense, an object can be either real, like the sadness of some person, or unreal, like an angel or a square. It can exist – an example of an existing object would be the Academy of Young Scholars and Artists in 2016 – or not exist, like the Academy of Young Scholars and Artists in 2005, which is also an object, but a non-existent one. To put it briefly, "everything which is not nothing, but which in some sense is 'something,' is an object" (Twardowski, 1977, p. 35).

Objects can be possible, and they can be impossible. An apple lying on the table in front of me is a possible object (and one existing in reality), while a square circle is an example of an impossible object. Objects are not limited to things – to assert that they are would surely be a questionable view, as things are only one sub-category of the category of objects. V(D)J recombination, or the creation of a stage character by an author, are not things, but they are objects; the same goes for the Stoic postulate of the premeditatedness of evil, mannerism, the measure of disorder, and many other objects about which the authors of texts included in this volume write.

The object of presentation is viewed *through* the content of presentation, in the act of presentation, where this is similar to the way a painter records a topic on a canvas *through* the picture, in the act of painting. Within this analogy, the topic corresponds with the object (e.g., the landscape), the picture corresponds with the content, and the act of painting corresponds with the act of presentation (cf. Twardowski, 1977, p. 13). The object of the presentation and the content of the presentation are not the same. One difference between them, for example, is that they can exist in different ways. The presentational content of an ideal or non-existing object is not ideal or non-existing. The content of the presentation of a non-existing object, should it be presented, exists independently of the non-existence of the object to which it pertains.

Object and content are to be discussed as the object and content of a particular presentation. Therefore, if we think only about some object, it will each time be the object of a certain possible presentation. Something similar goes for the content: if we discuss some specific content, that will be the content of a particular possible presentation. However, to render things more concise (in places where doing so will not engender inconsistency), we will use the terms *object* and *content* with the additional specification that they are the object or the content of a specific presentation. The context will decide whether the topic is one and the same presentation or, perhaps, many presentations.

3 Material Constituents and the Form of the Object

Anything which can be distinguished in an object independently of the very mode of distinguishing we call its part. Parts of a smartphone are a processor, a battery, a touch screen, etc. Parts of that type may be specified differently (albeit more narrowly) as pieces or parts in the usual sense. Nevertheless, in a smartphone, we can also discern that the battery cover covers the battery, that the touch screen is on the other side of the battery cover. In this way, parts of another type are distinguished: namely, relations between parts construed in the usual sense.

According to Twardowski (1977, p. 46ff), all of the parts of an object in the usual sense go to make up the *material* (German: *Stoff*) of an object, while all of the parts in the sense of relations distinguished in an object go to make up the object's *form*. Obviously, the material of an object is so strongly interwoven by different relations with the object's form that, in many cases, it is impossible to separate these effectively. An object's material and form are abstracta of that object, and even though they are initially inseparable, to distinguish them is to do so in thought.[3]

4 The Material Constituents of the Object

The material (or matter) of an object is the totality of its parts in the usual sense of the term "part". Parts are divided into simple ones, such as those that do not have parts, and complex ones, in which further parts can be distinguished. A *simple* object is an object that does not have parts; a *complex* object is an object with parts. The object's parts, viewed in terms of their mereological order, may be located within it more proximally or more distally: the more proximal part of a smartphone is its touch screen, while its most distal part is the technology from which it was made – for example, its capacitive sensing screen technology. The fact that a part is a more proximal or more distal one furnishes the basis for distinguishing the *order of parts* (Twardowski, 1977, p. 47). First-order parts are parts of an object, but their parts are second-order parts, parts of second-order parts are third-order ones, etc. Breaking down an object into mereological orders (degrees or levels) is, of course, relative. A kilo may be broken down into 100 decagrams or 1000 grams: it is hard to say which parts of a kilo are more proximal and which more distal if we are not apprised of the particular rule for carrying out such a procedure.

The order of the parts is just one of the rules for dividing something up into parts of a whole. The next rule of division is the way in which parts can be parts of an object (Twardowski, 1977, p. 48). As parts of particular objects, duration and extent may be their parts only in *one* way. The Academy of Young Scholars and Artists cannot have duration in three different ways – it was established and lasts, so as long as it exists, it will last. At the same time, the redness of a red pen is a part of that pen in a different way than redness conceived as a part of a spectrum's red band.[4]

The third way of dividing an object, alongside order and being a part in one or multiple ways, is division in terms of the dependencies pertaining to the existence of the object's parts (Twardowski, 1977, pp. 48, 49).[5] The latter are often tied to one another existentially: i.e. they are entangled in such a way that if one part exists, then the other has to exist too. That is the case with the redness and the extent of the apple lying on a table about which I wrote earlier: its redness cannot exist without the extent of its surface, and *vice versa*. The surface cannot exist without a specific colour. Colour in that sense is connected with extent, and there is an unbreakable relation of coexistence.

5 The Object's Form

The object's matter, as we have seen, is already incredibly complicated. However, the true complexity of an object is revealed in its form: that is, in the relations in which an object and its parts are involved. Following Twardowski, these relations are called the "formal constituents" of an object. We should also emphasize that Twardowski treats the relations of an object with other objects as a part of that object.

Among the formal parts, Twardowski (1977, p. 50) distinguishes two types: primary and secondary. (Twardowski speaks about "primary formal constituents" and "secondary formal constituents".) The *primary* formal parts are the relations between the parts and the object as a whole, while *secondary* formal parts are relations obtaining among the particular parts themselves.

Relationships between the parts of an object and the object as a whole come in two kinds. Once the parts create a whole, and the whole binds them together (for example, causally) and encompasses them, they can be considered relationships in the *strict* sense; meanwhile, at other times the parts can be just somewhat similar to a whole, or looked at separately may be smaller/larger than the whole, in which case Twardowski considers such relationships to be primary formal constituents only in an *extended* sense. Parts of a given whole, to be capable of being similar to the whole to some degree, must first be its constituents.

Primary formal constituents in the strict sense stand in an ontological relationship of *founding* with primary formal parts in the extended sense, and such founding itself constitutes a further level of multiplying relations. Because the relata of the founding level are relations (primary formal constituents), it can be said that they are secondary-level relations. Relations between relations of the second level will be those of the third level, etc.[6] In the case of some objects that analysis could go on endlessly. Such indefinite entangling of formal constituents is the key to answering the question about the core of the relation that binds the parts into a whole (see Twardowski, 1977, p. 56).

Formal secondary constituents of an object are those whose relata are the particular parts of an object, but not the object as a whole. The type of secondary formal constituents depends on its relata. When the relata are the material parts of an object, the secondary formal constituents are spatial relations between the material parts of, say, a smartphone.

The relata of these relationships may also be the primary formal constituents: i.e. the relations between the parts and the totality of the object. The differences between the ways in which the totality possesses the material parts can be compared as part and parcel of the analysis of the relations between the primary formal constituents. What is also to be taken into consideration as part of that analysis is the causal dependency obtaining between the different relations of an object and the object as a whole. If a smartphone has the right amount of RAM and a fast processor, it will also probably possess the feature of reacting quickly to the needs of a demanding user. We should add here that for Twardowski, the totality of secondary formal constituents, whose relata are primary formal constituents and those on the basis of which we can causally derive all remaining primary formal constituents in the discussed object, make up the *essence* [German: *Wesen*] of that object (Twardowski, 1977, p. 57).

6 Presentations Are Also Objects

The different types of parts of objects have been distinguished above in general terms. We should also point out that presentations can themselves be objects of presentations, as when I present a stage character right before the play. If such presentations are also objects, the types of parts distinguished above are also valid for them. Therefore, presentations may be broken down into material and formal parts, and it is possible to determine their further levels *ad infinitum*.

7 Adequacy

Twardowski writes:

> Just as the whole object is presented through a presentation, so the single parts of the object are presented through corresponding parts of the presentation. Now, the parts of the object of a presentation are again objects of presentation, and the latter in turn are parts of the whole presentation. Parts of the content of a presentation are contents, just as parts of an object are objects. In analogy to the way in which parts of an object form the whole uniform object, parts of a content form the complete content. (Twardowski, 1977, p. 39)

Since we are concerned with what it means to come to know an object, and the primary part of cognition is presenting an object, we are bound to be keenly interested in relationships between the contents of presentations and the objects of presentations. The object of a presentation is related to the presentation just because they are disclosed to us within one presentation. However, that aspect of their relationship does not have the same significance as how the presentations correspond with the content. To express it more precisely: in what manner do adequate parts of a presentation correspond with the object's parts (such that the particular object and that particular content refer to the same object), and *vice versa*? To be even more precise, our question is whether any formal and material part of the content of the presentation corresponds with a formal and material part of the object, and whether each formal and material part of the object corresponds with a formal and material part of the content.

That issue forms part of the classic philosophical problem of the essence of veracity. From amongst the products issuing from our cognition of an object, we can distinguish those that do and those that do not "hit" the object, those that do and those that do not correspond with it, those that describe the object as being how it is and those that describe it as how it is not, those that are harmonious with the object and those that are not, and those that are consistent with other presentations of it and those for which such consistency with others is not possible.

Each of the dichotomies listed above corresponds with a particular tradition of grasping the problem of truth. The analysis presented here pertains to the base level of the problem of veracity, because presentations, according to the view presupposed by our thesis, make up exclusively the basis of cognition. Moreover, cognition is understood here as the activity of an individual subject (even if it is slightly extended by, for example, online computational tools – see Smart, 2022), while the group aspect of cognition is omitted here (even though the results of scientific cognition have, to be sure, a social dimension – see Fleck, 1979; Hutchins, 2012), as is the theoretical aspect, in the sense that we are not aiming to deal with the question of *what it means for a particular scientific theory to be true.*

8 Addition and Subtraction of the Object's Parts: Simplification and Enrichment of the Presentation of the Object

Let us repeat our central question: it concerns whether each formal and material part of a presentation's content corresponds with a formal and material part of the object, and whether each formal and material part of the object corresponds with a formal and material part of the content of the presentation. That question, to be sure, is actually composed of many more detailed ones. Further levels of formal and material parts are discernible with each subsequent step. To successfully take the analysis in that direction, one would have to have appropriately complex and abstract structures at hand, which would model the structure of the content and the object (Wójtowicz & Skowron, 2022). We know that despite many attempts, such general structures have yet to be discovered. Below, though, we propose an overall topological understanding of the objects of interest to us.

Let us, then, try to answer that question not in an entirely general way, but on the basis of particular cases. Take, for instance, a concrete smartphone along with all its parts. Alongside those parts of a smartphone we have already mentioned, we should add the more complex relations it enters into with other objects: i.e. proximal spatial relationships, such as how it is located relative to other objects, and how it enters into network relationships with them. When it is connected to the Internet, it connects up with many other objects, including other smartphones, service providers' servers, internet websites, etc. The network of relations which it is involved in is incredibly varied, and having regard, perhaps, to the accumulative proliferation of orders of parts, endlessly rich. However, the question is whether, within the content of the presentation of a smartphone, examples of all the relationships just listed will be encountered? The answer to this is a negative one. When I present a smartphone to myself, I do not take into consideration all the relationships in which it is involved.

What is more, it is not just relationships that are omitted in the content of the presentation: I also do not consider its material parts, because I omit both duration as a part of a smartphone, as well as further mereological and temporal levels of its persistence, not thinking all the time about its beginning and end. I also omit many secondary formal components, such as the processor's location and the motherboard, which is seldom a component of the content of a smartphone's presentation.[7]

Until now, we have been talking about complex objects, in that we have distinguished their parts. The analysis up to now has not included cases of simple objects, that by definition do not have parts. Examples of simple objects are Leibniz's monads, a pure subject, or God. The content and object of presentation of simple objects are most probably related in the following way: that they are the content and object of one presentation. Do any other relationships obtain?

If we assume a proper breakdown into parts as described above, it would have to be said that there are no other relationships in force. This is because they cannot be said to be completely identical, as they are different objects. If anyone comes to know these objects at all, then this is not via a process having presentation as its basis. The importance of presentations decreases to the benefit of such intangible tools of cognition as intuition (for example, of the eidetic sort), and symbolic and abstract cognition.

Nevertheless, complex objects are often presented as simple objects.[8] In moments of strong subjective tension – for example, under stress, or being suddenly blinded by a bright flash – we tend to present complex objects as simple ones. The content of the presentation of a flash in a moment of surprise is not complex: we do not differentiate its parts, we do not analyze its colour or intensity – the content tends to be simple. States of tension or surprise are a violation of our standard mode of maintaining a certain poise, or relative peace and quiet, yet in those states a simplification of objects occurs. It might be conjectured that simplification is a cognitive strategy typical for subjects engaged in coming to know an object.[9] Cognition is, in large part, a matter of filtering what is available to the subject. It is impossible to contain in the content of a presentation all parts of the object of presentation. Attempts to fully specify an object, in which strict mutual adequacy between the parts of the content and the object is achieved, generally prove unsuccessful.

When we present complex objects as simple, we subtract some parts from them, and simplify them. A process that is the reverse of simplification is the conferral of additional internal diversity, which occurs *inter alia* in cases of optical illusion or vivid psychedelic hallucinations. The content of presentation of the image below includes more parts than the image itself. The grey spots which, when we look at the entire image, appear in the places where the white edges of the grid cross-sect, are a part of the presentation, but they are not a part of the object. This illusion is called the Herman grid illusion, and is one of many known illusions of that type.

Another example is furnished by the content of presentation of objects such as *a legal driver without a driving license*. Within the content of that presentation, it is natural to include *a driving license*. It is impossible to present to yourself *a legal driver without a driving license* without presenting some form of driving license. The driving license, although not part of the object, is a part of the object's presentation. There are many cases of such indubitable enrichment of objects in terms of content. More subtle examples are contributed by transcendental philosophy which, while examining the preconditions for our cognitive capacities, points to certain accessories needed by the subject

FIGURE 16.1 The Herman grid illusion (Source: http://en.wikipedia.org/wiki/Grid_illusion, Accessed: 06.04.2023)

(for example, sensual forms such as time and space), as well as some creative aspects of cognition involved in creating an object. However, pursuing a transcendental analysis would take us too far away from our main task here.

It can be the case that certain parts of the object do not correspond with any parts of the content, and *vice versa*. Although such a deficiency with respect to adequacy may have negative consequences, the idea that it wholly undermines the cognitive effectiveness of subjects engaged in coming to know an object would seem like an overly hasty conclusion. It seems that small differences pertaining to parts within the object should correspond with small differences pertaining to parts within the content of presentation – at least in respect of that presentation's "hitting" the object. The situation is similar with significant differences – with greater and more noticeable differences within the content – at least if assume that in a certain way we are, for example, asymptotically approximating to the object as we come to know it. If the relationships described here in terms of differences as to the content and the object do not make reference to the entire content and object, they certainly will obtain between selected parts of these.

Cognition exhibits a certain purposive character, which is that as we go through the process of coming to know something, we at the same time yearn for that which is being cognized to be accessible to us in full. Distortions, though undesirable, are an inherent component of the "adjusting" of the content to the object. They may be presumed to form a part of the conditions of possibility of cognition – in that the idea of a capacity for complete coherence with respect to cognition can, as we have seen, be quickly challenged.[10]

9 The Search for a Model of Object and Content[11]

As we have noted already, the problem of adjusting the model or the material to achieve adequacy as regards both the object and the content remains open. We know that despite many imperfections, correspondence takes place between parts of the content and parts of the object. In order to examine the nature of these correspondences, we must in some way imitate, or assign a structure to, such contents and objects. We know that there can be no relation of photographic mirroring between these structures, and also that in the process of presenting them many simplifications and diversifications will arise – as what, collectively, we call "distortions."

Within the content of its presentations, a subject somehow adjusts the parts to its properties and capabilities. It constructs certain presentations (cf. Jernajczyk,

2014) that amount to distortion, but which nevertheless do not become severed from the basic condition that a particular content should correspond with a particular object of presentation. Some, even if erratic, adequacy on the part of the content of the presentation still remains. Sometimes it is strict adequacy, while at other times it may be absent altogether. The strictness of this adequacy will consist in the following: that small changes in an object can be expected to compel equivalent small changes in the content, and *vice versa*. The contents of the presentation will be distorted, yet the process of distorting is not entirely unconstrained. It maintains "partial continuity," so to speak. So what does it consist in?

As an attempt at responding to the question concerning the structure of adequacy of the content's parts, we propose to model the structural aspect of both the content and the object with the help of topological spaces.[12] On such a basis, we believe that it will be possible to examine and model the nature of such distortion.

10 Dimensional Types Pertaining to Contents and Presentations

Assuming that certain topological spaces furnish a formal representation of the object of presentation and the content of presentation,[13] we may define the phenomenon of "partial" continuity of distortion in the following way.

Let F be the content, and G the object, respectively, of a certain presentation. We know that F and G are not identical in all respects. They may differ in, for example, the way they exist. However, an instance of full structural adequacy obtaining between them, should any such thing occur, would consist in their homeomorphism (topological identity): i.e. the existence of the bilaterally continuous and bijective function f: F⟼G. As we know, such adequacy will be too strong. Homeomorphism assumes continuous restructuring of the entire content to the entire object, and *vice versa*. Nevertheless, even if there are moments of discontinuity between the entirety of an object and the entirety of the content, as was mentioned earlier continuity may be maintained in certain places and on the level of parts. On that level, the concept of *types of dimension* proves helpful – one which Fréchet (who actually calls them, in French, *types de dimension*), defined as follows:

> Let F, G be two topological sets. F will be said to have a number (or type) of dimensions smaller than or equal to G, if F is homeomorphic with a part of G. This shall be denoted by
>
> $$dF \leq dG.$$

> Two sets H, K will be said to have the same type (or number) of dimensions if at the same time
>
> $$dH \leq dK; dK \leq dH.$$
>
> If $dH \leq dK$ but no part of H is homeomorphic with K, then H will be said to have a smaller number (or type) of dimensions than K: $dH < dK$. (Fréchet, 1928, p. 400)

The relation of *being-a-part* is expressed with the help of set theory – the relation of mutual containing between sets. Homeomorphic objects have the same types of dimension. However, the reverse phenomenon does not occur, and objects of the same type may not be homeomorphic. The elementary example is a closed interval (an interval with limit points) and a straight line. They are not homeomorphic. A straight line, contrary to the interval, is not compact. However, if we consider an interval included in the closed interval, but this time without limit points, i.e. an open interval, it is homeomorphic with the straight line. And going in the other direction: if we take a closed interval which is a part of a straight line, it is homeomorphic with the initial closed interval. The real straight line and the closed interval have the same type of dimension, although they are not homeomorphic.

To sum up: if object G is homeomorphic with a certain part of object H, and object H is not homeomorphic with any part of object G, we say that object G has a smaller type of dimension than object H, which we record as $dG < dH$. If, between any objects X and Y, there is neither $dX < dY$ nor either $dY < dX$ or $dX = dY$, then we say that X and Y are incomparable in terms of their dimensional types.

An example of a sequence of objects with a growing dimensional type is shown in Figure 16.2.

We see that the greater the dimensional type, the more complex the object. Simplification of objects, and the conferral of additional diversity upon them, relies in the content of presentations on ascribing a smaller type of dimension in the

FIGURE 16.2 A series of objects of progressively higher dimensional type

former case and an ever-higher type of dimension in the latter one. More precisely, if some part of the presentation's content has a smaller dimensional type than the corresponding part of the object of presentation, then we say that that part of the content of presentation is a simplification. Meanwhile, a conferral of additional diversity on a particular part takes place when a part of the presentation's content has a higher dimensional type than the corresponding part of the object of presentation. The content of presentation will be strongly simplificatory if each of its parts is a simplification in relation to each corresponding part of the object, and it will be strongly diversificatory if each part of that content amounts to a conferral of additional diversity in relation to the corresponding part of the object.

The actual complexity of cognition depends on the premises present in advance as part of what constitutes its basis: namely, that during a presentation some parts of it will simplify parts of the object, and some will confer additional diversity on them. Presentation is a process, and as such is extended over time. One object corresponds with sequences of presentations. During that process, one part of the object may have a lower type than the corresponding part of the content, and another may have a higher type than its counterpart.

It is worth noting that even if contents can have either a simplificatory or a diversificatory impact on an object, they maintain the basic relationship between content and object responsible for the fact that some contents correspond with that particular object, and that the object is presented through that particular content. That relationship is based on the comparability of the dimensional types of the corresponding parts: i.e. that we can say that the first one is higher than the second, the second higher than the first, or that they are equal. When such comparability as to types of dimension does not occur, the contents of presentations fall short of the object. Cognition does not "reach out" successfully to its object. This thesis is an expression of the philosophical conviction that the contents of presentations, even if they distort the object by subtracting or adding something, still in some basic sense refer to the object and to some extent present it. That is a kind of epistemological optimism.

In order to carry out – in a reliable and detailed manner – an analysis of objects' and contents' presentations under the mantle of topology, we would have to formulate appropriate postulates regarding adequacy relating to the types of parts differentiated earlier, and to the parts of topological spaces: ones similar in form to the postulates of quantum mechanics. However, due to the scale of such an undertaking, that would exceed the scope of this paper.

11 On the Union of Presentation and Object

Some parts of the content of presentations do not fully correspond with parts of the objects of presentations and – *vice versa* – some parts of objects of presentations are not fully represented in the contents of presentations. Despite this, objects are somehow cognized. Therefore, there must be some parts of an object whose equivalents are parts of the content of the presentation of that object. These parts, given that they are responsible for cognition's adequacy and its being "on target," hold a particular cognitive significance, and for that reason have been picked out using a different name. Twardowski calls them *characteristics* (German *Merkmale*). A characteristic of an object is that part of it, which is at the same time a part of the content of the presentation of that object. In the words of Immanuel Kant: "A characteristic is that about a thing which constitutes a part of the knowledge of it" (quoted in Twardowski, 1977, p. 79).

Twardowski also posed the question of whether there are parts of the object of presentation that every object possesses. Amongst the natural candidates for that type of part, we find identity and unity. Identity, however, need not be a part of the object of presentation: it occurs, for example, in the set of contrary objects. On the other hand, unity is a part of each object. Each object of presentation is a homogenous totality, even when it is a complex object: it is *one* object, and it is presented as *one* object. Therefore, unity exists in every content of the presentation, which is why it is a characteristic of every object (Twardowski, 1977, p. 86).

12 In Place of a Conclusion

The process of cognition is complex. Right up to this point, we have focused exclusively on a certain phase of cognition relating to the presentation of the cognized object. We have also assumed that a presentation is important because it is a founding and enabling moment of cognition. It is on the basis of presentations that we are able to carry on with cognition. What is presented can be judged – that is the essence of cognition, but it might be both desirable and repellent.

Each and every cognition, so long as it does claim to be cognition, should culminate in the formulating and pronouncing of a certain judgment. Yet even though cognition may not always achieve a definitive completion in such terms, this need not mean that it is not then cognition – it will often have a certain value or cognitive power, albeit of a somewhat unfinished and incomplete kind. Nevertheless, it is worth noting that the process of cognition, right from its beginnings, is in danger of being simplified or rendered more internally diverse many times over, where this may subsequently lead to further distortions. In particular, when such a process culminates in the formulating and issuing of a judgment, some distortions present at its outset may engender distortions at its ending. Being aware of such impairments (or inherent properties?) of cognition, especially in terms of the cognition of complex objects, will then permit an adequate adjustment of the strength of assertion of the judgment made.

Acknowledgements

Many thanks to Agata Hamilton for translating this text into English, and to Carl Humphries for proofreading the final version. The research was supported by research project No. 2017/27/B/ HS1/02830, financed by the National Science Centre (Poland).

Notes

1 The term *cognition* can refer to the act or process of cognition, or to the product of cognition construed as the result of that process. In subsequent parts of this text I do not distinguish on each occasion between these two meanings, as it is clear from the contexts in which sense the term is being used.

2 The English term *presentation* corresponds to the German *Vorstellung* and Polish *przedstawienie*.

3 Perhaps the fullest study of the *form* and *matter* of the object is to be found in the analysis carried out by Ingarden; see (Ingarden, 2016, Chapter VII).

4 An interesting formal case study in which one and the same object is a part of another object in three different ways was presented by Thomas Mormann (Mormann, 2009).

5 Twardowski here speaks rather about parts of some content, and not parts of objects, because objects may not exist. Nevertheless, we will stay with objects: the dependency pertaining to existence might be understood in a wider sense – such as, for example, dependency of a kind consisting in necessary coexistence.

6 This phenomenon can be formally examined in many ways. One of the abstract ways to do that is by introducing what in mathematical category theory is called "higher category theory". A category consists of objects and adequately specified arrows (morphisms) between objects. An example of objects would be groups which, along with group homeomorphisms as arrows, create a *group* category, i.e. the 1-category. If you consider arrows between the arrows, that leads to the 2-category, etc. (see Baez, 1997).

7 A wider analysis of perceptual simplifications, and the omission of parts of objects from the contents of eyewitness presentations, may be found in (Jernajczyk, 2013). Jernajczyk develops a thesis to the effect that perception is discrete – that it is similar to a wide-meshed sieve through which only an interrupted and discontinuous representation of the world seeps. We will not attempt here to discuss the continuity/discreteness of perception (and the object of perception, which is also discussed by Jernajczyk); however, we do think that adequacy in some way entitles us to assert continuity – to put it in Jernajczyk's words – between the world and its representation. "Lossy" representations of reality are not, as one might suppose, "lossy" in an unconstrained way. If we lose something from an object, then this can only occur in a way that means that we do not lose the object itself – assuming that we are targeting the object with our cognition.

8 In the analysis of simple objects we must omit relationships in which these objects are involved, should they themselves be involved in any relationships at all. Otherwise no object would be able to fit the definition of a simple object.

9 In research and scientific practice, that judgment has been substantiated. In many texts included in this volume, scholars point out the importance of making things simple, such that it might well be assumed that simplification is an unavoidable phase of the process of scientific cognition. For instance, M. Rowińska-Żyrek points out how important the scheme "simple question – complicated analysis – simplification – simple answer" is in researching *Helicobacter pylori*. That scheme bears the traces of a cognitive strategy. See also (Jernajczyk, 2014).

10 We know that the claim that a complete, absolute cognition is impossible is a powerful one. Indeed, in the case of complex objects it seems be valid. Simple objects, on the other hand, as long as they allow for the subject to approximate to them with its cognitive tools, would seem to leave room for two outcomes only: either complete inaccessibility and absence of cognition, or full cognition.

11 Readers may wish to skip over this part of the article on a first reading.

12 Such modeling, although it might seem inappropriate, is essentially based on the approach to modeling quantum phenomena in quantum mechanics. To be more precise, the core of this type of modeling is included in the postulates of quantum mechanics. The first postulate says that each quantum system corresponds with a Hilbert space, while the second says that for each observable there will be a self-adjoined operator operating in the Hilbert space related to the quantum system on which we are carrying out the measurement. Similar to that, we want to say that in the sense defined above there are corresponding topological spaces. At the same time, we propose to model the entire universe with the *top* category: i.e. the category of all topological spaces, in which arrows are continuous transformations. For an answer to the question of why such attempts at mathematical modeling in metaphysics are worth pursuing, see (Wójtowicz and Skowron, 2022).

13 For other uses of topology in philosophy, see Lewin (1936), Fine (1995), Schulte and Juhl (1996), and Kaczmarek (2019a, 2019b). For a book-length review of applications of topology in ontology (that calls such applications "topo-ontology"), see Skowron, 2021.

References

Baez, J. (1997). "An Introduction to n-Categories." In E. Moggi & G. Rosolini (Eds.), *Category theory and computer science, lecture notes in computer Science* (Vol. 1290). Springer-Verlag, Berlin. arXiv:q-alg/9705009v1.

Burkhardt, H., & Dufour, C. A. (1991). "Part/Whole I: History." In H. Burkhardt & B. Smith (Eds.), *Handbook of metaphysics and ontology* (pp. 663–675). Philosophia Verlag.

Fine, K. (1995). "Part-whole." In B. Smith & D. W. Smith (Eds.), *The Cambridge companion to Husserl* (pp. 463–486). Cambridge University Press. https://doi.org/10.1017/CCOL0521430232.011

Fleck, L. (1979). *Genesis and development of a scientific fact*. The University of Chicago Press.

Fréchet, M. (1924, August 11–16). "Number of dimensions of an abstract set." In J. C. Fields (Ed.), *Proceedings of the international mathematical congress held in Toronto* (pp. 399–413), Toronto: University of Toronto Press. mathunion.org/fileadmin/ICM/Proceedings/ICM1924.1/ICM1924.1. ocr.pdf

Hutchins, E. (2012). Concepts in practice as sources of order. *Mind Culture and Activity, 19*(3), 314–323. https://doi.org/10.1080/10749039.2012.694006

Ingarden, R. (1972). *O sądzie kategorycznym i jego roli w poznaniu*. In R. Ingarden (Ed.), *Z teorii języka i filozoficznych podstaw logiki* (pp. 222–259). PWN: Warsaw.

Ingarden, R. (2013). *Controversy over the existence of the world* (Vol. I, A. Szylewicz, Trans.). Peter Lang.

Ingarden, R. (2016). *Controversy over the existence of the world* (Vol. II, A. Szylewicz, Trans.). Peter Lang.

Jernajczyk, J. (2014). Lossy representations of excessive reality. In Ł. Huculak, B. Skowron, K. Dąbrowska, J. Jernajczyk, M. Zakrzewska, & R. Zarzycki (Eds.), *Excesss and lack* [*Nadmiar i Brak*]. Wroclaw.

Kaczmarek, J. (2019a). Ontology in the tractatus logico-philosophicus: A topological approach. In G. M. Mras, P. Weingartner, & B. Ritter (Eds.), *Philosophy of logic and mathematics: Proceedings of the 41st International Ludwig Wittgenstein Symposium* (pp. 397–414). Berlin, Boston: De Gruyter. https://doi.org/10.1515/9783110657883-024

Kaczmarek, J. (2019b). On the topological modelling of ontological objects: Substance in the monadology. In B. Skowron (Ed.), *Contemporary polish ontology*. Berlin: De Gruyter. https://doi.org/10.1515/9783110669411-009

Lewin, K. (1936). *Principles of topological psychology*. McGraw-Hill Book Company.

Mormann, T. (2009). Updating classical mereology. In C. Glymour, D. Westerstahl, & W. Wang (Eds.), *Logic, methodology and philosophy of science. Proceedings of the 13th International Congress*. King's College.

Schulte, O., & Juhl, C. (1996). Topology as epistemology. *The Monist, 79*, 141–147. https://doi.org/10.5840/monist19967916

Smart, P. (2022). Minds in the metaverse: Extended cognition meets mixed reality. *Philosophy of Technology, 35*, 87. https://doi.org/10.1007/s13347-022-00580-w

Skowron, B. (2021), *Część i całość. W stronę topontologii* [*Part and Whole: Towards Topo-Ontology*], Oficyna Wydawnicza PW. PDF: https://philarchive.org/archive/SKOCIC-2

Twardowski, K. (1977). *On the content and object of presentations. A psychological investigation,* translation and introduction by R. Grossmann, The Hague: Martinus Nijhoff. (Originally: *Zur Lehre vom Inhalt und Gegenstand der Vorstellungen. Eine psychologische Untersuchun*, Vienna 1894).

Wójtowicz, K., & Skowron, B. (2022). A metaphysical foundation for mathematical philosophy. *Synthese, 200*, 299. https://doi.org/10.1007/s11229-022-03760-5

CHAPTER 17

The Complexity and Multifacetedness of Conversation

Karol Striżyk, Piotr Kruk, Justyna Ziółkowska, Bartłomiej Skowron and Magdalena Witkowicz

Abstract

Conversation, as an integral part of our everyday life, is a subject of research in various fields. In this paper we aim to look at the different layers of conversation through multidisciplinary lenses. First, by describing two distinct speech disorders resulting from brain damage, we highlight the connection between brain functioning and communication. Language usage can be profoundly disrupted when particular brain areas are affected. Second, shifting to a sociopsychological perspective, we look into potential reasons behind common misunderstandings. Conversations are shaped by social scripts, impacting communication patterns and intercultural dialogues. Then we delve into the logic of conversation, focusing on the mechanism of presupposition. Through the analysis of simple interactions, we unveil the subtle nuances of conversation and the role presupposition may play in manipulation. Finally, we explore the transformative potential of dialogue in psychotherapy. Conversations serve as powerful tools for treatment, transcending specific techniques. They can have a profound impact on brain functioning and well-being. Our exploration underscores the multifaceted nature of conversation and the complex effects of this seemingly simple process.

Keywords

conversation – speech disorders – logic of conversation – conditions of conversation

1 Introduction

Conversation is an ineliminable element of our everyday lives. We talk at home, at work, at school, in the street, about trivial and important matters, sometimes in private, sometimes on business. For most of us, it is as natural an activity as breathing – we engage in it without giving it any thought. So why is it that philosophers, logicians, psychologists, cognitive scientists, linguists, brain scientists, biologists and other scientists have studied (and continue to study!) conversation? How is it that this element of our everyday life can be disturbed in various ways under different conditions? Why has there been such an intensive development of psychotherapy, the essence of which is conversation? What can logicians add to our knowledge about conversation?

In this short article we will try to show that, despite its apparent simplicity, conversation is a complex process, conditioned not only by normal brain functioning but also by logical and socio-cultural rules. To this end, we will draw on work in psychology, but will also look into other fields that deal with this phenomenon. We will start with brain disorders affecting the mechanisms of speech, then move on to social, cultural and logical aspects of verbal communication, before ending by making reference to clinical psychology and psychotherapy.

2 Biological Bases of Speech Disorders

The development of knowledge about the brain is largely a history of diseases, tragic accidents and sometimes physicians' mistakes. The neurological mechanisms of speech production and comprehension are no different. We will briefly describe two typical examples of disorders resulting from brain damage, in which the ability to use language is disrupted in various manners.

More than 150 years ago, Paul Broca, a French surgeon, met a man in the hospital who was referred to as Tan (Dronkers et al., 2007). His real name was Leborgne and his nickname came from the fact that, although with different modulations, "tan" was the only word he spoke. However, he seemed to understand what was being said to him. After Tan's death, Broca performed an autopsy and found extensive damage to the posterior part of the left inferior frontal gyrus. Within a short time, similar lesions were seen in around a dozen more patients. The affected area of the brain now bears the surgeon's name, and the disorder resulting from the damage is called Broca's aphasia. Today, it is recognised that this type of aphasia is associated with speech difficulties, while speech comprehension remains almost intact (Dingwall, 2005). This has significant implications for conversation:

"I: now tell me what this thing was with your legs last week or week before. (2 sec.)

R: a: no good a: (4 sec.) [sighs] a: (1 sec.) ache (2 sec.) and (5 sec.) a: (4 sec.) a: (8 sec.) a: (2 sec.) knees (1 sec.) and ankles (3 sec.) a: (5 sec.)

I: what did you do about it (4 sec.).

R: a: home (3 sec.) doctor (2 sec.) and (1 sec.) legs (1 sec.) m: a (4 sec.) a: (2 sec.) walking (2 sec.) no good.

I: no good huh?"[1]

The above extract is an excerpt from a conversation with a man suffering from this disorder.[2] It illustrates the speech disturbances commonly occurring when Broca's area is damaged – lack of fluency of speech, difficulties with grammatical sentence construction and a limited vocabulary.

Although the above extract can be considered a typical example of expressive aphasia, in reality, the severity of the disorder can vary greatly from person to person: from a complete loss of the ability to speak (as in Leborgne's case), through levels similar to the one demonstrated above, to relatively minor problems with finding the right word.

At the time of Paul Broca's work, neurologist Carl Wernicke was studying the area where the auditory nerve and the auditory cortex meet (Dingwall, 2005). He found that damage to this area also affects speech, although in a very different way than in Broca's aphasia (and again, both the disorder and the brain area were named after the discoverer). In the case of Wernicke's aphasia, impaired language comprehension is considered its main characteristic. People suffering from this disorder do not necessarily lose fluency of speech, its syntax may also be preserved. However, in their utterances, they increasingly deviate from the topic or answer a completely different question than the one asked. In a variant of this disorder, so many twisted and non-existent words (neologisms) are used in speech that its content becomes completely unintelligible. Interestingly, people suffering from Wernicke's aphasia may not realise that there is something strange about the things they are saying. This phenomenon – lack of awareness of one's disease – is nowadays called anosognosia.

The consequences of various types of lesions (brain damage) are the most striking examples of how complex and unstable our ability to communicate verbally is. However, they are relatively rare, hence we do not experience problems related to them on a daily basis. Much more frequently and subtly we are affected by the consequences of the psychological processes of social cognition during communication.

3 Socio-Cultural Conditions of Conversations

Most of us are able to describe the steps needed to start and drive a car. On a similar basis, we can describe the schematic course of a chess game or a field hockey match. According to some social psychologists, we are able to create similar descriptions for typical social situations. Such schemas in psychology are called scripts and are defined as mental representations of events and actions or sequences of them (Schank & Abelson, 1977). They are formed on the basis of acquired knowledge and experience related to a given type of situation. It follows that although there is a certain script of a certain situation that is more or less consistent for a particular society, the script of the same situation may vary for two different people. For example, a person only familiar with the formal rules of hockey will most likely be surprised to see a "mosh pit" of players punching each other on the ice. On the other hand, to a person who has had the opportunity to watch hockey matches, a "mosh pit" will not seem unusual. The latter one knows that in the social sphere, there is some consent to players' fights, as they potentially belong to the script of a hockey match.

Importantly, the overwhelming majority of our interactions with other people are not excluded from the scripted categories. "A visit to the office," "shopping at a discount grocery store," "a police check," "meeting a buddy." Each of these scripts allows us to engage in these types of situations without having to devote as many cognitive resources to them each time, both during the planning and the course of the events. Also, in the case of formal contacts, the scripts make them more efficient and less stressful; we then use a limited repertoire of possible questions and answers. As long as the conversation follows a script, much of the action and reaction is semi-automatic. We are aware of what we are doing, but we do not think carefully about every single word. It is worth mentioning that information processing is then shallower, hence, according to researchers, some of the methods of social manipulation are based on activating scripts.

A good example of intuitive awareness and the usage of scripts is a method that appeared some time ago among some of the individuals asking for change on the streets of Wroclaw. The method involves starting a conversation by saying "I'll be honest. I'm collecting change for beer." Someone has rightly pointed out that people have become accustomed to and have automated their (usually negative) responses to requests containing made-up stories about both the motivation and the situation of the asker. The asker's unexpected sincerity brings the element of novelty necessary to jolt the bothered out of their automatism and at least increase the chance of having a real (sic!) conversation. That is, of course, until the asked notice a pattern, learn it and master their response as part of the "sincere beggar" script. Interestingly, this method can also derive from automatisms, according to which "honesty" is a positive trait to be appreciated and rewarded with help. And all this in a few words!

As a consequence of hypothetical scripts, a number of problematic areas occur, where communication may not go our way. First, we may be participating in a situation for the first time and we lack adequate knowledge of it. The notorious "visiting the future in-laws" definitely falls into this category. What should be said, what can they ask about and how not to look like a fool? Subsequent visits usually fall into the category of less stressful ones, as we (at least partially) know what to expect and we gain some kind of scheme – a script. It is also possible that our script of the situation we are in is not adequate for it for some reason. This problem usually occurs when we tend to overgeneralise and apply our knowledge of previous events to others, without taking the changing circumstances into account, in particular – people and their impact on the course of events. This leads to the third and, in our opinion, the most interesting of all the potential issues. Sometimes, when two people who interact with each other perceive the situation they are in differently, they may execute entirely different scripts. This is often the case when one of the dialogue participants "puts on an act" in the relationship by hiding their real intentions. It will also happen when individuals misread each other's intentions. This often results in an abrupt breakdown when our "almost-spouse" turns out to have only collegial feelings for us, and the "grandson" turns out to be a money extortionist.

Incompatibility of scripts and conventions of conversation can also happen when people from two different cultures meet. The easiest way to show the differences is to compare the extreme cases. Beyond doubt, Iran and Germany are separated by an enormous distance – both geographically and culturally. Certain consequences of this distance are shown in a study that analysed the beginning of telephone conversations (Taleghani-Nikazm, 2002). Thus, Iranians, once they identify their interlocutor, move on to a series of questions about their well-being and health. As in the United States (for example), the response is a quick and brief "okay" (or, rather, the Persian equivalent), and then the "how are you" is reciprocated. However, this is not the end of the exchange of pleasantries. In a similar manner, inquiries about spouses, children or, in general, members of the immediate family are made – even in situations where the phone is not answered by the person we were trying to reach. Germans rarely engage in this ritual exchange of "how are you/how is your family." Once the participants recognise each other, they usually move on to the reason for the call. If the well-being questions are asked, they elaborate on them instead of moving on after a brief 'okay' or 'fine.' The questions about family do not come up either. The study is interesting as it not only juxtaposed two cultures and showed the differences between them, but also analysed conversations between the representatives of the two nations. Thus, Iranians, who have lived in Germany for many years and are fluent in the German language, when starting conversations automatically impose a pattern of questions about well-being and family, implemented from their culture. In one of the recorded excerpts, the German interlocutor breaks the convention of short, positive answers to the "how are you" question and, in line with his understanding of the situation, considers these inquiries an invitation to elaborate on the matter. In another fragment, there are "disruptions" in the conversation (relatively long pauses) after questions about well-being – which may be due to the incompatibility of the 'starting the conversation' script. There are many more examples in the study similar to the ones described.

It may appear that the described research looks into completely trivial and insignificant characteristics of conversation. Nothing could be further from the truth. In the first half of the 20th century, Bronislaw Malinowski pointed this out, referring to the role played in communication by short and often pointless utterances about, for example, the weather, insignificant events, and obvious matters. He stated that these small talks help "establish the bonds of personal connection between people gathered in one place" (Malinowski, 1972). He called the realisation of this task the phatic function of communication. Moreover, some researchers have suggested that these seemingly trivial elements of conversation may have been the link in the evolution of apes to today's *Homo Sapiens* (Dunbar, 2004). What was the hypothetical course of this transformation? Gathering in large groups, on the one hand, facilitates survival (thanks to, among other things, a greater ability to fend off predators), but on the other hand, it exposes them to dangers coming from other individuals from the same herd and requires cooperation and coordination of activities. According to the researchers, the mechanism that primates developed to maintain appropriate social bonds was grooming. During this activity, both the grooming and the groomed release endorphins that strengthen relationships in the herd. However, the effectiveness of grooming is limited – with other duties (such as obtaining food), the time that can be spent on it allows for the formation of a group of fifty individuals on average. And here, according to this concept, is to appear conversation, which, due to its characteristics (conversation is not limited to two participants, it can also be performed with other activities) allows to maintain more relationships in less time. Consequently, it allows to form a more numerous group. According to this hypothesis, language and the need to orient oneself in increasingly complex societies were the reasons for the development of human intelligence. In light of the above, are small talks trivial? Perhaps, but they have a special meaning for humans.

Adopting the assumption of simultaneous development of language and human communities allows us to look at intergender differences in linguistic communication in a different way. Intergender differences in the tasks performed, which already existed in the pre-linguistic stage of development of *Homo Sapiens'* ancestors, may have been reflected in the forming language and the ways in which it was used. Moreover, this approach also implies that these pre-linguistic intergender differences in functioning, which were "imprinted" in the language, will be replicated through it and transmitted to subsequent generations. Because of the complexity of this process and its significant consequences for almost every sphere of human life, it is a frequent object of interest for a wide range of communication researchers in various contexts.

In one study of mothers' verbal interactions with their daughters and sons, the researchers managed to show that girls are asked more questions, mothers repeat girls' texts more often than boys' texts, and the average length of statements directed to girls is longer (4.4 to 3.7 words). On the other hand, commands and demands are formulated more often towards boys. The frequency of parents' conversations with their children about emotions – especially sadness – is also different. These conversations are more often conducted with girls, regardless of the gender of the parent. The same research showed that dialogues about emotions, if conducted with girls, are constructed in a way that emphasises their interpersonal nature. All these factors have a profound impact on formal differences in language use between adults of different genders. According to the study, men use longer words on average, use the pronoun "I" more often, use the present tense more often, make more grammatical errors, tend to use active voice, evaluative adjectives and references to the so-called social rightness ("most people," "everyone knows that."). In contrast, women's speech typically consists of more rhetorical questions, longer sentences, verbs specifying or referring to movement, adverb intensifiers, the use of oppositions and adverbial sentence starters. All these differences translate into very specific communication dilemmas affecting our daily functioning. One study tested communication efficiency in heterosexual dyads. Its results suggest that women more commonly take a passive position in interactions, especially towards men. They are also more likely to experience problems in interactions when the situation requires them to turn the passive position into an active, cooperative one and take part in solving a problem. In a similar way, the previously mentioned differences concerning sadness may result in different coping strategies in adulthood. In turn, greater ease in communicating and talking about emotions, as well as locating them in interpersonal space, may, by contrast, lead to women's different construction of various key concepts, such as love. This can lead to radically different behaviours and expectations that we have of loving relationships.

4 The Logic of Conversation – The Mechanism of Presupposition

The complexity and multifaceted nature of conversation is also the subject of research in modern logic (Budzyńska et al., 2015; Szymanek, 2001). Among the many issues addressed, we will draw attention to the logical mechanism of presupposition. It is included in many everyday conversations, is distinguished by the fact that it has a clearly delineated practical aspect and is also relatively easy to articulate.

Let us return to the previously mentioned example of visiting the in-laws. Let's say that this is already the second consecutive visit, it is no longer accompanied by stress, because the script of such a visit has been established. Let's assume that we are talking about Jack and Agatha, and that Jack is at Agatha's parents' place. Let's also assume that during a conversation over dinner, Jack's father-in-law asks him a question:

1. Have you Stopped Cheating on my Daughter Yet?

This is a resolution type of question, i.e. the correct answer to this question should be "yes" or "no". Nevertheless, surprised Jacek is in a difficult situation, because no matter what he answers, it comes out that he cheated on Agatha. If he answers "yes," it means that he cheated on Agatha, while if he answers "no," it means that he continues to cheat on her.

Note that if Jack's father-in-law had asked in a slightly different manner, i.e., if he had posed the following question:

2. Have you Not Stopped Cheating on my Daughter Yet?

Then Jacek would find himself in an equally difficult position.

Leaving aside Jack and his father-in-law, let us try to find the logical mechanism hidden in this example. For simplicity's sake, let's use an affirmative sentence:

3. He Stopped Cheating on Her.

And let us ask, does sentence (3) make logical sense, i.e., is it true or false? Sentence (3) is false if he still cheats on her, while it is true if he has cheated on her and no longer does so. If we assume that he has never cheated on her and is still not cheating on her, then sentence (3) can be neither true nor false. Therefore, the veracity of the sentence:

4. He Cheats on her,

is a necessary condition for the sentence (3) to have a logical value, i.e. to be true or false (true and false are two core logical values). In this case and in similar ones, we will say that sentence (4) is a presupposition of sentence (3).

Generally speaking, we say that a sentence *P* is a presupposition of a sentence *Z* if the sentence *Z is* determined as to its logical value if, and only if, *P* is true. Assuming that sentences with a certain logical value, i.e., being true or false, are meaningful sentences, we can say in short that *P* is a presupposition of *Z* if the meaningfulness of *Z is* equivalent to the truthfulness of *P*.

Note that Jack's father-in-law could have posed both question (1) and its (certain) negation (2), and their presupposition would still be sentence (4). Let's give a simpler example. Sentence:

5. Jarosław Kaczyński's Wife is a Beautiful Woman;

As well as its denial:

6. Jarosław Kaczyński's Wife is Not a Beautiful Woman.

We see that sentences (5) and (6) have one and the same presupposition, i.e., the sentence:

7. Jarosław Kaczyński Has a Wife.

If (7) were false, then both sentences (5) and (6) would have no logical value. Sentences (5) and (6) share the same presupposition, i.e., sentence (7). This general characteristic of presupposition is called invariance under negation.

Implicit in the relatively simple question of Agatha's father is a certain relationship as to logical value. This dependency is not very complex, nevertheless, the presented logical analysis is a certain beginning – often simple dialogue mechanisms lead to hidden and complex logical constructs such as dialogical logics, an accessible discussion of which can be found in: Budzyńska et al., 2015.

In Jack's defense, let's add that in response to such a dishonest question (2), he can simply cancel the presupposition, stating that he has never cheated on Agatha, and the question is asked with a dishonestly concealed (if in fact he did not cheat on Agatha) presupposition. Jack's example is somewhat exaggerated in order to make the reader more alert to the presence of presupposition in everyday conversations, particularly in view of the fact that presupposition is also one of the logical mechanisms of manipulation. Indeed, phrases like, "Our success is the result of years of effort," "Don't stress," "Don't chase money," "He keeps his agreements," "He broke his vows," can be part of manipulation. However, presupposition has many uses in a conversation, for example, it can shorten it, as in the following example:

A. Did you enjoy the film?

B. It's a film like any other.

The answer (B) can contain many pre-judgments, which by no means always have to be directly spoken in a dialogue situation. After all, answer (B) in the right situational context can mean:

(B') I didn't like the movie, it was boring and I don't want to watch any more movies made by this producer.

5 Treating with Conversation

Until now, we have mainly referred to private communication between people. However, it is the professional conversations that reveal additional layers of complexity of human communication. We will use conversations in the clinical work of psychologists and psychotherapists as examples.

In the psychological, especially psychotherapeutic practice conversation serves as a working tool. Through conversation, the psychotherapist obtains information about patients and builds a therapeutic relationship with them. Despite the apparent simplicity of asking questions and listening to answers – usually viewed as a natural ability that does not require perfecting – the multitude of tasks incumbent on psychotherapists who conduct conversations makes them the most difficult part of clinical work. Their complexity has so far failed to be put into foolproof procedures, and conducting therapeutic conversations is still largely seen as an art (MacKinnon et al., 2006).

Despite the long history of therapeutic interactions based on conversations, until recently, the reflection on the power of conversation was mainly theoretical, derived from declarations made by patients. One of the particular elements of the recent scientific debate on psychotherapy was the emphasis on the distinctiveness and competitiveness of various psychotherapeutic modalities and approaches. Cognitive-behavioural psychotherapy was contrasted with psychodynamic or humanistic-existential psychotherapy. However, this way of thinking – that the effectiveness of therapy is determined by the specific methods or therapeutic techniques – has increasingly been challenged by research indicating the role of engagement in therapy by both the therapist and the patient and the role of the relationship built through conversation (Norcross & Lambert, 2019a, 2019b).

Research findings indicate that the conversation and the relationship established through it are important factors influencing the process of therapeutic change (e.g., Hill & Norcross, 2023). Therapeutic success depends more on the quality of the therapeutic conversation than on the techniques used to work with the patient. This proves the power of conversation.

However, the results of studies on the neurobiological mechanisms of psychotherapy are of particular interest to us, as they demonstrate the power of conversation even more evidently. They show that professional therapeutic talk alters the electrophysiological activity, biochemistry, blood flow and structure of patients' brains, which paves the way for changes in the behaviour and the way the patients experience life (Barsaglini et al., 2014). However, despite disagreement among researchers and reservations about the accuracy of brain imaging methods, a link between observed brain changes and the process of psychotherapy has been systematically demonstrated. Moreover, the magnitude of observed changes in the brain under the influence of psychotherapy is comparable to the effects achieved with pharmacotherapy or electroconvulsive therapy (Linden, 2006).

For example, it has been demonstrated that patients with excessive activity in the prefrontal cortex, which is typical for depression, experienced a decrease in this activity after a series of interpersonal psychotherapy sessions. The effect was similar to that achieved with pharmacotherapy (Brody et al., 2001). Another study has shown that in patients with a diagnosis of chronic fatigue syndrome, whose brains are characterized by a decrease in grey matter volume in the prefrontal cortex, gray matter volume increased after 16 sessions of cognitive-behavioural psychotherapy (de Lange et al., 2008).

Therefore, it turns out that therapeutic conversation has as much power to influence our brain as pharmacotherapy. And, while the correct brain function we referred to at the beginning of the chapter determines the possibility of having conversations, therapeutic conversations also have the power to influence the brain and, as a consequence, change our lives. The seeming simplicity of conversation translates into its complex and still not fully recognized effects.

Notes

1 The letter 'I' stands for a person carrying on a conversation, while 'R' stands for a man suffering from Broca's aphasia. A colon after a voice indicates a prolonged sound. The length of the pause between statements is indicated in parentheses. A period and a question mark indicate that the passage was spoken with a descending or ascending tonality, respectively.

2 The cited conversation is an excerpt from a video found on youtube at https://www.youtube.com/watch?v=f2IiMEbMnPM. For other recordings in which similar problems are presented, search under 'aphasia' (e.g., 'Broca's aphasia' or 'Wernicke's aphasia').

References

Barbara, G. (2008). Gender differences in the verbal expression of love schema. *Sex Roles*, *58*(11–12), 814–821.

Barsaglini, A., Sartori, G., Benetti, S., Pettersson-Yeo, W., & Mechelli, A. (2014). The effects of psychotherapy on brain function: A systematic and critical review. *Progress in Neurobiology*, *114*, 1–14.

Brody, A. L., Saxena, S., Stoessel, P., Gillies, L. A., Fairbanks, L. A., Alborzian, S., et al. (2001). Regional brain metabolic changes in patients with major depression treated with either paroxetine or interpersonal therapy: Preliminary findings. *Archives of General Psychiatry*, *58*(7), 631–640.

Budzyńska K., Kacprzak M., & Yaskorska O. (2015). *Dynamika dialogów w ujęciu formalnym* [*Dynamics of dialogues in formal terms*]. Warsaw: Publishing House of IFiS PAN.

Coupland, J., Coupland, N., & Robinson, J. D. (1992). "How are you?": Negotiating phatic communion. *Language in Society*, *21*(2), 207–230.

De Lange, F. P., Koers, A., Kalkman, J. S., Bleijenberg, G., Hagoort, P., Van der Meer, J. W., & Toni, I. (2008). Increase in prefrontal cortical volume following cognitive behavioral therapy in patients with chronic fatigue syndrome. *Brain*, *131*(8), 2172–2180.

Dingwall, W. O. (2005). The biological basis of human communicative behavior. In J. B. Gleason & N. B. Ratner (Eds.), *Psycholinguistics* (pp. 65–120). Gdańsk: Gdańsk Psychological Publishing House.

Dronkers, N. F., Plaisant, O., Iba-Zizen, M. T., & Cabanis, E. A. (2007). Paul Broca's historic cases: High resolution MR imaging of the brains of Leborgne and Lelong. *Brain*, *130*(5), 1432–1441.

Dunbar, R. I. (2004). Gossip in evolutionary perspective. *Review of General Psychology*, *8*(2), 100.

Eisenmann, B. (1997). Gender differences in early mother-child interactions: Talking about an imminent event. *Discourse Processes*, *24*(2–3), 309–335.

Fivush, R., Brotman, M. A., Buckner, J. P., & Goodman, S. H. (2000). Gender differences in parent-child emotion narratives. *Sex Roles*, *42*(3–4), 233–253.

Hill, C. E., & Norcross, J. C. (Eds.). (2023). *Psychotherapy skills and methods that work*. New York, NY: Oxford University Press.

Linden, D. J. (2006). How psychotherapy changes the brain – the contribution of functional neuroimaging. *Molecular Psychiatry*, *11*, 528–538.

MacKinnon, R. A., Michels, R., & Buckley, P. J. (2009). *The psychiatric interview in clinical practice*. Washington, DC: American Psychiatric Publishing.

Malinowski, B. (1972). Phatic communion. In J. Laver & S. Hutcheson (Eds.), *Communication in face-to-face interaction* (pp. 146–152). Harmondsworth: Penguin.

Norcross, J. C., & Lambert, B. E. (Eds.). (2019a). *Psychotherapy relationships that work: Volume 1: Evidence-based therapist contributions.* New York, NY: Oxford University Press.

Norcross, J. C., & Lambert, B. E. (Eds.). (2019b). *Psychotherapy relationships that work: Volume 2: Evidence-based therapist responsiveness.* New York, NY: Oxford University Press.

Schank, R. C., & Abelson, R. P. (1977). *Scripts, plans, goals, and understanding: An inquiry into human knowledge structures*. Hilsdale: Erlbau.

Szymanek K. (2001). *Sztuka argumentacji. Słownik terminologiczny* [*The art of argumentation. A glossary of terminology*]. Warsaw: PWN.

Taleghani-Nikazm, C. (2002). A conversation analytic study of telephone conversation openings between native and nonnative speakers. *Journal of Pragmatics*, *34*(12), 1807–1832.

CHAPTER 18

Algorithmic Truth Theory (ATT)

Roland Zarzycki

Abstract

This article sets out to analyse what a fact is, and how the simplicity behind this notion has recently been challenged by the emerging phenomenon of post-truth. With reference to the new social-media virtual reality, in which the reliability of information is backed by the quantity, not quality, of its sources, it compares the standard approaches to truth (the correspondence ToT, coherentism, the redundancy ToT, the pragmatic ToT). Information bubbles, propaganda bots, and crowd opinions have become so natural in the modern world of the internet that the concept of a fact seems to require reconsideration. It must be confronted with the blurring of the foundations of our current knowledge and science, insofar as these are affected by commodification and the uncontrolled growth of information. An attempt is made to pinpoint the main mechanisms underlying these processes, and to provide a simple model explaining the differences between past, current and expected future perceptions of facts. On this basis, it is claimed that the complexity and interdependence of information are gradually increasing, and that coherentism is therefore emerging as the narrative that models perceptions of truth by various e-communities most accurately. Thus, it can be argued that the perception of facts will soon require unprecedented and extraordinary skills of a sort that we need to start developing.

Keywords

truth theories – post-truth – commodification of truth – critical thinking – power of algorithms – artificial intelligence – algorithmic capitalism

1 Towards the Multi-Dimensional Complexity of Information or How Does Post-Truth Challenge the Simplicity of Facts?

In the 20th century the issue of the social nature of reality resounded with full force, expressed perhaps in the simplest way in a bold statement of

Wittgenstein (2021) "the limits of my language mean the limits of my world." It was then strengthened and multiplied by a varied array of research conducted in the fields of philosophy of language (Austin, 1975), literary studies (Rorty, 1989), anthropology (Whorf, 2012), sociology (Berger, & Luckmann, 1971) and others. The baseline intuition underlying this concept is that the world seems to be what it seems to be, that certain objects, such as a hundred-yuan bill, are significant in our lives, because of the construction of the symbolic universes that we create by the semantic power of our languages. While many would agree that this is indeed the case with the banknote, it seems much more controversial to say that the existence of the sun depends on our language.

In a somewhat surprising (for the contemporary fast-thinking world) way, the old philosophical questions of existence, of truth and of language are not only interdependent, but also intertwined with everyday life. This was interestingly illustrated with an example of the 09/11 terrorist attack(s) on the World Trade Center provided by Pinker (2007). A huge controversy comes from a seemingly simple question if this event should be considered one single or two separate attacks. Even though it might seem like philosophical hair-splitting, Pinker convincingly argues that the exact opposite is true: it is a purely pragmatic, legal, and painfully material issue, with a clear interpretation expressible in American dollars. Depending on whether the attack on the World Trade Center was a single event or two separate events, the value of the insurance and subsequent financial compensation for the owner should be single or doubled. The in-depth importance of language for the architecture of reality is manifested in this basic picture.

Another fundamentally relevant observation was brought to our attention by Fricker (2007) together with her concept of epistemic injustice. From epistemological perspective, we can observe, following Fricker, that until the concept of "sexual harassment" was formed, victims of sexual harassment found it difficult to conceptualize their problem, not even mentioning expressing it in a convincing way.

Nevertheless, though still questioned, discussed and controversial, the thesis on the linguistic-social construction of reality,[1] signifies an important shift from the XIX-century intuitions of Kant to an understanding of the language–reality nexus by contemporary sociologists. However, the XXI century brings yet another transformation. A significant and growing share of our life activity actually moves to cyberspace, which is controlled in a broad sense by algorithms. These, in turn, although created by man, contribute to the emergence of phenomena that seem to elude human understanding and supervision.

In consequence, parallel to the change in the very understanding of what reality is and after recognizing the cognitive limitations of mankind, it is the concept of truth that requires thorough reconsideration. For even if contemporary science provides us with more and more accurate models, descriptions, and predictions of reality, as long as the reality is linguistically-socially constructed, we remain dependent on human hermeneutics in any attempt to make these descriptions and predictions meaningful (Glynos, & Howarth, 2007). In other words, language and reality are consubstantial, and truthfulness is just another aspect of "explaining the world" language games.

In front of the paradigm change as well as ontological challenges related to the overwhelming role of algorithms, the meaning of truth calls for reinterpretation. The growing dependence of our cognitive schemes on the algorithm-driven digital sphere, especially due to the recent rapid development of Artificial Intelligence (AI), prompts the next phase of reflection on the nature of reality itself. The fact that some negligence has already taken place in this regard, can be easily spotted by observing the spreading phenomenon of post-truth, which significantly reduces the quality of the human condition (Fuller, 2018; Fridlund, 2020). The transition to algorithmically constructed realities is the main point of interest of this paper. Our main goal is to conceptualise how mechanisms describing these new ontological-epistemological tensions impact our traditional approaches to truth. In the following sections of the paper, we describe in what sense the complexity of reality increases. We then explain how this increasing complexity inevitably triggers changes in the architecture of power and leads to an ever deeper dependence of the individual on structural factors. In the last section we formulate a first outline of the Algorithmic Truth Theory (ATT) that provides a narrative to effectively and pragmatically describe the changes we are subject to.

2 Complexity as a Discursive Trick

Typically physicists convince us of the existence of an objective world that they apparently observe and describe by means of mathematical models and physical experiments. Due to the fact that the mediation of this process of cognition through language is out of the question, phenomena appear to be *simple* if their descriptions can be simplified by the application of mathematical instruments and they appear *complex* otherwise. Moreover, there is a specific academic discipline, the theory of complexity, which explains the meta-phenomenon of complexity. Within this paradigm complexity becomes a resultant of temporality and non-linearity characterising a given process as argued by Coveney and Highfield (1995). At the same time, however, these models created by physicists themselves generate tremendous complexity that is literally projected onto the world. This peculiar *Ptolemaic syndrome* provides vivid examples of how our perceptions of complexity are subordinated to the structures of discourse, and thus also to authority and power.

Within the materialistic approach, complexity might be also expressed in terms of unpredictability of the cumulative effect of actions of individual agents, which leads to a spontaneous emergence of complex, organised structures, even though each one of the atomic actions was governed by simple behavioural rules (Kauffman, 2002). Yet another, and much older conceptualization comes from Kolmogorov (so called *Kolmogorov complexity*), where the complexity of an object is defined by the length of the shortest computer program (in a given language) that generates the object as output (cf.: Chaitin, 1987).

Kind of surprisingly, these considerations lead to the ideas of algorithmic entropy and freedom. It is significant that the increase in the complexity of the model meant to explain a given phenomenon typically and structurally evokes introduction of a new, external narrative element semantically equivalent to freedom. One might recognise this phenomenon as a link between the practices of modern scientists and language spells used by the Maori people (Lévi-Strauss, 1950). The theoretical position developed by Penrose in *The Emperor's New Mind* within an old, XX-century debate on the non-algorithmic complexity of human consciousness and the possibilities of quantum computers can serve as an interesting illustration of this meta-methodological mechanism (Penrose, 1989).

On the other hand let us consider the following example: a patient goes to the doctor and learns that his incomprehensible and disturbingly complex set of ailments has a *proper* name, let it be "conjunctivitis". Instantly the phenomenon becomes simple. Not only in the sense that s/he is given a prescription and a few simple manoeuvres reduce the ailment but also in the sense that the whole sensation starts appearing to be simple. It can be observed here how naming kept its magical power over reality (cf.: Bodenhorn, & Vom Bruck, 2006).

The question of complexity can be explained according to alternative cognitive traditions functioning in the scientific discourse (cf.: Mainzer, 1994). Referring to the semiotic model of Peirce we could apply the category of complexity (and simplicity, respectively) to the sign (the scientific model adopted), to the very object of our cognition, or to the ramification of the sign. Even such a *simple* procedure would introduce more transparency into the nature of complexity, which is of great importance for further understanding of the structure of truth. What appears to be complex or simple within a given sign system and its ramification may not be like that after we change the system or ramification method. This understanding of complexity allows one to recognize that the qualification of a given model as the complex is a matter of discourse.

Accordingly, this is yet another dimension in which discourse impacts the perception of truth. In the following sections, we will see how this perception is affected by complexity.

3 Complexity as Unpredictability

The development of culture and technology has always led to more and more sophisticated techniques of oppressing the body and soul. Rituals and fantasies armed with new tools affect the psychology of individuals and the sociology of interpersonal relations. The more complex these technological tools are, the more difficult it is to realise these consequences of their application. Somewhere in the future, however, it quite often translates to destructive practices, from mutilation of bodies, through plagues of depression, to the destruction of self and others. An emancipated wo/man, liberated from the bindings of his/er biology, employs technology to dominate nature, being at the same time rarely interested in critically assessing his/her own power (Adorno, & Horkheimer, 1944; McPhee, 1989; Agamben, 1998). It can be also easily observed how this effect has only been amplified by the development of the culture of narcissism underlying every individual empowerment (Lasch, 2018).

The complexity of technological tools, which were put into mass use in the XX century, already exceeded the human ability to predict even the mid-terms effects of their application. The scale of imprudence in satisfying momentary human desires, fueled by rampant consumerism, turned complexity into a key source of the affliction of the XXI century wo/man. This unpredictability is fundamental to all areas of daily existence, from the sphere of intimate relationships, through the professional sphere, to the very understanding of reality and the meaning of life.

One illustration of this phenomenon is the complexity of sexual entanglement with digital pornography. The spread of video pornography at the end of the XX century had a highly negative impact on interpersonal sexual relationships (Fagan, 2009). As a whole lot of research proves, digital solutions lead to a gradual degradation of the sexual life at an alarming rate (cf.: Lemma, 2021). Current research on sexual interaction in digital contexts (SIDC) shows three basic patterns of using technology in sexual life: (1) mediation in arranging physical sexual interaction; (2) realising sexual interaction in digital space; (3) getting sexual satisfaction from a machine (Brand et al., 2021). The scope and intensity of these practices show how more complex and sophisticated technology, that remains at our disposal and addresses our desires, creates less and less predictable environments.

Love and belonging can serve as another example of a basic human need (Maslow, 1943) affected by complexity. The digital world clearly addresses it, but in quite a specific way, based on mechanisms oriented on competitiveness, collecting and ostentation embedded in social media platforms. Hungry for relationships, internet users instinctively search for other human beings and emotional reinforcements such as attention, touch, care (Agger, 2015). This turns love and belonging into a highly lucrative business, which translates to purely financial profits of the intermediaries and investors (cf.: Seebach, 2017).

Unfortunately, the complexity of relationships in which we function in the digital world, driven by the need to increase our recognisability and measured by likes, reachout and number of friends, makes it difficult to assess any real quality of these relationships. What is even worse, this complexity prevents an individual from giving enough attention to any of these digital contacts to build a satisfying relationship. The structure of human bodies and minds allows them to pay attention to dozens of individuals, but when they have to pay attention to thousands, it turns out to be insufficient. Hence, typically, users of social networks only receive a digital semblance of closeness, which is not followed by human touch, leaving their bodies in a state of deprivation. And although modern technology is working fast to cover this demand too, it will not happen that quickly (Farroni et al., 2021). Due to the complexity of these relations, the overall impact and long-term effect of such a substitution seem to be difficult to predict. In this context, it is also not surprising that so many people in developed economies choose to escape into nature, seeking refuge from the complexity in the backwoods or mountain cottages (cf.: Karsten, 2020).

Regardless of the current shortcomings of technology, the market's attempt to replace family members, friends and peers with digital influencers is a matter of time. In this context, it is worth asking already today whether the meaning of true love is going to be defined by algorithms. Not in the sense of a textbook definition, but the actual practices of teenagers who lived their youthful loves with digital substitutes.[2]

4 Overwhelmed by the Complexity

Another instance of complexity creeping into daily lives is the number of legal contracts people are forced to sign and regulations they need to accept. Clearly, one does not sign these obligations under duress, but even if it is not direct coercion, it is certainly structural coercion. Without these mass approvals, any successful functioning in modern societies would be massively hampered.

Statistically speaking, the documents are being accepted thoughtlessly. A Norwegian consumer agency campaign found that the average Norwegian uses 33 apps on their smartphone (BBC 2016). During a provocation event released on the occasion of the publication of the research report, users read aloud the terms and conditions of all these apps, which together run longer than the New Testament.

Apps' terms and conditions are just the tip of the iceberg. Another widely known absurdity is the need to accept cookies every single time an internet user enters a new website. This irrational regulation, enforced by the European Union, allegedly for the security of the users, is actually a purely cynical act of privatising responsibility (Zarzycki, 2021). Interestingly, at the current stage blind-signing has become a deeply internalised common practice and an open secret at the same time. Lawyers and governments turn their blind eyes to this fact, but actually being a responsible citizen and consumer would require spending the greater part of our lives reading contracts and bylaws that govern our actions.[3]

On a macro scale, contemporary civilization seems to be unable to cope with the consequences of the complexity of the economic and legal systems it has created (cf.: Rodrigues, 2020). With the current size of the global market, royalty regulations drive stratification and create negative socioeconomic impacts. Similar global leverages apply to incomes of famous athletes or music stars, sport events producers, media stations and music labels, that massively contribute to the growing scale of global inequalities (cf.: Piketty, 2014). The debate about unfair pay gaps, however, is always diluted by the out-of-control complexity of the global system (Lowrey, 2019; Leicht, 2020). Even if this situation is not universally accepted (Kelly-Smith 2002), mere dissatisfaction does not necessarily mean that the mechanisms underlying the injustice and inadequacy of the current system have been properly recognized or addressed. Moreover, even within the framework of the scientific discourse, there is a consistent attempt to separate economic and ethical threads (Wheeler, 2019, presumably for the sake of simplicity).

On the other hand, as argued by Goldfarb (2006), it might seem that politics depends solely on the decisions of the big players. This picture is simple but flawed or rather flawed because oversimplified. The constellation of millions of tiny conversations at the *kitchen tables* all over the world, where families and friends talk about politics is too complex to enable envisioning its overall effect and significance. This does not mean, however, that these millions of chats are not important. As argued in The Politics of Small Things, eventually they do change the world, as was the case with the great political transformations of the nineties in Central and Eastern Europe (Goldfarb, 2006). If Goldfarb is right, then the issue of grasping the complexity of these phenomena would become fundamentally important for the condition of XXI-century democracies in yet another sense.

5 Machines and Power

In the digital world of information overload, those who have the capacity and tools appropriate to employ this wealth will undoubtedly increase their advantage over the rest. More precisely, those who will be able to create sense architectures that transform amorphous information into meaningful narratives, will gain control over the discourse and truth (Zarzycki, 2013), which is basically just an XXI century update of Foucault's insights on the power–knowledge nexus (Foucault, 1980; cf.: Aberra, 2012). Geopolitically speaking, access to data is of no value unless there is sufficient hardware and software at the disposal of a given state or entity to transform the data into information, then into knowledge, and finally into wisdom (UNCTAD, 2021).

As the report on the digital economy by UNCTAD (2021) shows, while we already observe serious economic stratification in the world, the digital stratification following (and clearly, reinforcing) the former one is even more striking. Countries without the access to appropriate infrastructure and competence, are powerless when facing exabytes of available data that mean nothing. Thus the digital stratification is visible not only at the level of internal divisions within a given society, but to an even greater extent on the scale of global differences among the states. In terms of the capacity to exploit and benefit from the data-driven economy, the advantage of China and the United States is huge and growing (UNCTAD, 2021). As we read in the report,

> together, they account for half the world's hyperscale data centres, the highest rates of 5G adoption in the world, 94 percent of all funding of AI start-ups in the past five years, 70 percent of the world's top AI researchers, and almost 90 per cent of the market capitalization of the world's largest digital platforms. (UNCTAD, 2021)

Moreover, the largest digital platforms originating from these two countries are massively investing in building their technological advantage, especially in the area of AI. As UNCTAD (2021) notices, "they have become global digital corporations with planetary reach, huge financial, market and technology power, and control over large swathes of data about their users". As it is also emphasised, it is precisely due to this asymmetry that we need a global governance of data and cross-border data flow regulations, which would guarantee equal share of all the states in the profits of global digital development (UNCTAD,

2021). At the same time, however, it seems that the attempt to regulate this area as well as recent desperate actions aimed at limiting AI in connection with data privacy concerns (McCallum, 2023) is contrary to the particular interests of the main actors, and therefore also to the current state of political will.

It is precisely due to the fact that the current complexity of the digital processes exceeds by far the available technological tools which could help simplify it, that states, societies and individuals are dependent on publicly available algorithms offered by the digital giants, mainly from China and the US. Coming back to the practice of naming, this phenomenon also received its name, which, not without a reason, indicates its ideological origin. The concept of *dataism* applies to all these contexts, where the "information flow" becomes the "supreme value" (Brooks, 2013). Again one could argue that it is just another update to Foucault's general theory of knowledge-power (Foucault, 2008).

Success in the modern world requires a specific conversion of financial capital into a highly specialised technological capital, and more precisely, specific digital means of production, which make one capable of conducting structural analyses of large data sets, thus unlocking the above mentioned potential of digital wealth. In a natural way, market processes catalysed accumulation of these new forms of capital, and, in turn, the emergence of the power elites and netocracy, as diagnosed two decades ago by Bard and Söderqvist (2002). In this context one can also recognize the inner transformations of capitalism itself, where power starts being exercised by intelligent algorithms or rather owners of these algorithms (Kreft, 2018). The sense of helplessness in front of the global complexity of the digital sphere is shared by many, starting from ordinary surfers, through influencers, specialists, to companies and governments. Moreover, it is structure-wise necessary that any possible doubts in this regard have to remain ignored, as claimed by agnotological capitalism (Betancourt, 2016). Taking into account how limited is our understanding of the way these algorithms operate, especially those of them which are black-box-based, their actual power remains unnoticed (Beer, 2009).

It is worth noting that algorithms are already being used on a large scale within truth-determination protocols. For example human semi-slave labour aimed at controlling fake news is now being gradually automated by means of intelligent algorithms (Kreft, 2018). The political problem is, however, that contemporary fashionable fact-checking algorithms are not only written by specific people, representing specific views of their employers, but also owned by these employers having their own business and political interests. Filtering of information, under the guise of ensuring security, leads to uncontrolled spread of censorship. This in turn can have even more disastrous consequences for the quality of communication than letting people navigate on the waves of information on their own. One of the reasons supporting this approach is that the rationality of digital giants does not always go hand in hand with the needs of the democratic world order, hence it is not about the technological efficiency of the algorithms but rather about the political control and liberty (cf.: Bilić, 2018).

Eventually, an intriguing illustration of this type of discourse control abuse were the policies adopted by various digital actors that decided to censor the discussions on the recent COVID-19 epidemic. Any inconsistencies with the mainline narrative of undisputed rightness of mass vaccinations promoted by selected experts were brutally removed from many social media platforms. Clearly, it is beyond the expertise of this paper to analyse who was right in this discussion, but what is significant is that any attempts to discuss the topic have been ruthlessly suppressed, which deserves special attention. Moreover, this unprecedented and anti-democratic digital silencing campaign was implemented with the consent of all major actors, including civic societies. One could reasonably argue that COVID-19 proved that discourse and truth control in the free world can happen at any time, not even triggering any serious resistance.

6 Conversations with ChatGPT

One of the cutting-edge research techniques, gaining recently a lot of recognition and popularity, is AI-supported research, which is facilitated by the availability of open source free software, ChatGDP for instance. What is significant in this context is that even though ChatGPT is a relatively transparent algorithm, the very complexity of its operations makes it not fully understandable for a human being, which in turn can be perceived as one the generic moments of losing critical control over information processed by the AI. Some insights into the role of intelligent algorithms for the functioning of reality and truth emerge from the following conversation which was conducted by the author (ChatGPT, 2023).

Q1: What is the role of algorithms and intelligent algorithms in the contemporary world for determining what is true as well as people's beliefs about what is real and what is not?

ChatGPT: Algorithms and intelligent algorithms play a significant role in shaping people's beliefs about what is real and what is not in the contemporary world. With the vast amount of information available on the internet, it can be challenging to discern what is true and what is not, and algorithms are increasingly being used to assist with this task.

Intelligent algorithms are designed to learn and adapt to new information, which makes them particularly useful for determining what is true. They can analyze[4] large datasets and identify patterns, which can help to distinguish between real information and fake news or misinformation. These algorithms can also help to identify biases in data and sources, which can help to ensure that people's beliefs are based on accurate information.

However, algorithms can also perpetuate biases and reinforce existing beliefs if they are not designed and trained correctly. If algorithms are trained on biased data or are programmed with biased rules, they can perpetuate those biases and further entrench false beliefs. It is therefore essential to ensure that algorithms are designed and trained with fairness and accuracy in mind.

Overall, the role of algorithms and intelligent algorithms in shaping people's beliefs about what is real and what is not is complex and multifaceted. While they can be useful tools for determining truth and combating misinformation, they must be used with caution and care to ensure that they do not perpetuate biases or reinforce false beliefs.

Q2: Would you agree that in the same sense that Berger and Luckman claim that reality is socially constructed, today's reality is algorithmically constructed?

ChatGPT: Yes, I would agree that today's reality is algorithmically constructed to a certain extent. In the same way that Berger and Luckman argued that reality is socially constructed, algorithms and intelligent algorithms are increasingly playing a role in shaping our perceptions of reality.

Algorithms are used in a wide range of applications, from social media feeds and search engines to financial markets and medical diagnoses. They are designed to analyze data and make decisions based on that data, which can have a significant impact on how we perceive the world around us.

For example, the algorithms used by social media platforms to determine what content to show users can create filter bubbles, where people are only exposed to information that reinforces their existing beliefs. This can create a distorted view of reality, where people are only exposed to information that confirms their biases and are not exposed to alternative perspectives.

Similarly, algorithms used in financial markets can create feedback loops that amplify market trends, leading to a distorted view of the economy and market conditions.

Overall, algorithms are playing an increasingly important role in constructing our perception of reality. While they can be useful tools for analyzing data and making decisions, it is important to recognize their limitations and potential biases to ensure that they do not distort our understanding of the world around us.

Q3: how can a human control whether the mechanisms accompanying the aggregation, segregation and processing of information by intelligent algorithms do not lead to distortion of the image of reality, taking into account the extent to which many people uncritically use automatically generated intelligent suggestions?

ChatGPT: Controlling the mechanisms accompanying the aggregation, segregation, and processing of information by intelligent algorithms to prevent the distortion of the image of reality is a complex task. However, there are several steps that can be taken to minimize the risk of distortion:

1. Ensure transparency: One way to control the impact of intelligent algorithms is to make their decision-making processes more transparent. This could involve requiring companies to disclose the algorithms they use and how they make decisions, allowing people to understand how the information they receive is being curated.
2. Encourage critical thinking: It is important to educate people about the limitations of intelligent algorithms and encourage critical thinking about the information they receive. This could involve providing resources for fact-checking and teaching media literacy skills.
3. Develop ethical guidelines: The development of ethical guidelines for the use of intelligent algorithms could help to minimize the risk of distortion. These guidelines could be developed by industry organizations, academic institutions, or government agencies and could outline best practices for the use of intelligent algorithms.
4. Promote diversity: The biases of intelligent algorithms can be reduced by promoting diversity in the development and training of algorithms. This could involve increasing the diversity of the teams that develop and test algorithms and ensuring that algorithms are trained on diverse datasets.
5. Encourage algorithmic accountability: Ensuring that there is accountability for the decisions made by intelligent algorithms can help to prevent distortion. This could involve creating mechanisms for people to challenge decisions made by algorithms and for companies to explain how decisions were made.

Overall, controlling the mechanisms accompanying the aggregation, segregation, and processing of information by intelligent algorithms requires a multi-faceted approach that involves transparency, education, ethics, diversity, and accountability.

Considering how mechanically ChatGPT simulates being intelligent (Wolfram, 2023) on the one hand, and understanding how saturated it is with stereotypical, human-like thinking on the other (Airoldi, 2022), we should not be surprised, that answers to the above questions are politically correct, polite and apparently reasonable. ChatGPT does not provide us with anything new regarding the discussion on algorithmically constructed reality and truth, because its thinking seems to be anchored in what is known and safe. What is particularly interesting, however, is the algorithm's *belief* that there is some truth that it can discover through "pattern identification". Thus one of the main conclusions from this discussion is that ChatGPT did not read Berger and Luckmann carefully enough. Secondly, ChatGPT believes that truth is about recognizing patterns.

7 The ATT

The methods, which people apply to judge what is true, are often characterised by referring to some of the popular theories of truth, including the correspondence theory of truth, the coherence theory of truth and the pragmatic theory of truth. At the same time, as already stated, the development of technology and the increase in the complexity of communication games make it difficult to apply even the most basic truth criteria, apparently making these well-known theories less and less reliable.

Though sceptics warned us long ago that our senses can deceive us, the level of difficulty in distinguishing truth from appearance has never been as high as in the era of deep fakes. The ability to imitate a recording, behaviour, voice of a human being, but also the application of much more prosaic disinformation methods, prevent individuals from efficiently verifying if a given statement is true by determining its connection to reality, thus making the correspondence theory of truth obsolete. Modern wo/man encounters similar problems when trying to apply the coherence theory of truth. A multitude of opinions generated by bots and trolls of various sorts and origins, information bubbles and echo chambers, big data that cannot be easily analysed without advanced skills and tools, dedicated portals or research institutes financed by large interest groups are just examples of phenomena that have a negative impact on the truth verification process. Even the pragmatic theory of truth, which hypothetically exempts an individual from the necessity of external verification, making it subordinate to the internal interest of the individual, becomes extremely difficult to apply in the VUCA world. Liquid modernity, which calls into question the very meaning of existence, in conjunction with the growing technical unpredictability of actions due to the high complexity of social, economic and communication structures, make pragmatic approach an increasingly ambiguous criterion.

Clearly, in practice people do not follow any strict guidelines of any particular theory. Human behaviour can be described in accordance with the adaptationist theory of truth, which Godler (2020) defines as creative manoeuvring between various more and less accurate beliefs, information and sources, depending on available resources, in particular time. Although it is journalistic practice that is characterised in this way by Godler, there are many indications that it can be adopted to accurately describe activities of an average person.

It seems that the complexity of the surrounding world does not encourage people to explore the sources of information (Hagmann et al., 2020), who rather prefer to remain in a state of active ignorance. The fact that it is becoming more and more difficult for the people to determine the veracity of the incoming information does not mobilise them to increase their efforts in this area, but rather triggers resignation. This coincides with the general findings of Betancourt (2016), which indicate that this type of ignorance is a structural precondition for maintaining the current global market system. The feeling of not being able to determine the truth, as well as the feeling of the consequences of using unverified or incorrect information, fosters dishonesty and deception attitudes (Keyes, 2004). Importantly, at an individual level, this also contributes to the spread of the attitude of indifference to truth, which is the essence of the phenomenon of post-truth (Gudonis, 2017).

On the other hand, liberal societies have ingrained needs for freedom and civic agency based on some sort of determinable truth. For these reasons, among others, the activities of Cambridge Analytica, which offered services in the area of increasing the effectiveness of political persuasion, aroused so much controversy, indignation and accusations of manipulation. It seems that the general opinion accepts a situation in which citizens err and their sources of opinion are selective, as long as this erring is not pre-organised by someone. The difficulty here is that the massive outburst of fact-checking initiatives we are currently noticing in response to the transformation of truth structures seems to ignore the fact that people's views are based not on facts but on opinions. Beliefs and worldviews result from narratives that are convincing for a given person, and the interpretation of the truthfulness of facts is subordinated to these preferred narratives. Ultimate truth as an unshakable point of reference for our actions should be interpreted rather in terms of self-fantasies of mankind.

The coexistence of people and machines changes the perception schemes of both parties, but also the structure of expectations of human agents, especially in circumstances of information overload, which cannot be easily coped with by the people for both technical and emotional reasons (Andrejevic, 2013; Carr, 2014). On the one hand, the algorithmic logic impacts society in terms of social integration (Ninno, & Totaro, 2014), on the other hand, Airoldi (2022) clearly shows why this phenomenon should be perceived as a progressively reinforcing feedback loop. Clearly, the significance of the economic-structural context of these processes would be difficult to overestimate. As authors such as Eubanks and Zuboff show, what we observe is a kind of modern version of class conflict elevated to its digital incarnation. This is obviously reflected in the capitalist-consumer orientation of the contemporary intelligent machines that are programmed and trained according to a presupposition structure appropriate to the neoliberal worldview framework (cf.: Mager, 2012).

The growing technological capabilities of these machines in the area of control, supervision and verification of truthfulness run parallel to the decline of exactly the same competences on the part of human beings. Thus it seems natural that these human beings choose to rely on the effects of the technology they themselves have developed over the centuries. Current trend towards variously understood transhumanism emerges as a similarly natural phenomenon. On the wave of a peculiar mixture of enthusiasm, indolence and something that was identified by Broussard (2018) as *technochauvinism*, control over sense-making processes is transferred to machines. And although a lot has been already said about the raising power of algorithms, it is the extension of this power to this particular sense-making dimension that marks the most fundamental concession, because, according to Berger and Luckman, it is ontological. In this sense, in the first decades of the XXI century, we observe a shift from the social creation of reality to its algorithmic creation.

Finally, the concept of the algorithmic theory of truth can be formulated: to be true (and real) is to be a result/indication produced by one of the algorithms that have formerly received a popular mandate to provide people with reality. Clearly, the power of algorithms in this area will be the greatest within the digital sphere, but the interdependency of the biological and digital realities will easily transfer the senses created within the latter onto the structure of the former.

8 Conclusions – Make It Simple

Within a constant process human beings expand their vision, deepen understanding, create senses, and thus contribute to the increasing complexity of the real, and simultaneously they strive to simplify it to offer themselves a sense of epistemic relief and ontological security. Language is being impoverished and philosophy is pushed aside for the sake of simplicity and short-term efficiency, which tends to become the only currency that counts in a market-driven world. The point when machines and algorithms are powerful enough to take the burden of caring for truth and reality has been reached. Therefore it can be observed, how people gradually get rid of these inconvenient responsibilities for the price of giving up control over the creation of senses of their own world. This newly created algorithmic reality begins to live its own life and generate its own meanings, the complexity of which is so much beyond human cognitive capabilities. The dependence on the truth determined algorithmically becomes more and more explicit and technological assistance seems to be unavoidable. In the next step humans will presumably learn to interpret the world in a way that will prevent them from suffering cognitive dissonances resulting from the potential mismatches between their own observations and the indications of the algorithms or other digital authorities.

Perhaps the most significant aspect of these changes is that algorithms are not independent entities originating from a different world order, but rather economically controlled tools. In practice, this means that behind the technological enthusiasm there is a ruthless structure of power. The ideals of a free democratic world do not allow a free man to be controlled. However, in a world full of algorithms controlled by institutions, companies or more affluent members of society, coexisting with human beings in a non-distinguishable way and on equal communication terms, the discourse and truth are going to be controlled on an unprecedented scale.

This dialectical tension between simplicity and complexity demands to be overcome. Is it possible for the things essential to human existence to remain simple? Could individuals be granted the right to simplicity in accordance with the postulates of epistemic justice? Interestingly, even intelligent algorithms do not cope equally well with all dimensions of reality. Critical thinking, which is a competence of breaking patterns, seems to be particularly resistant to algorithmization. Thus it might be crucially important that human criticality remains a key element of the criteria of truthfulness to which we subordinate our reality.

This text is conceptual and paradigmatic, which constitutes its natural limitation. However, if the phenomena and mechanisms identified in it would be considered significant, the investigation is certainly worth being continued and followed by more specific research, including empirical studies of the complexity-truth nexus.

Notes

1 Not necessarily in the specific sense of the Whorf-Sapir thesis.

2 As in a slightly different context was brilliantly illustrated by the movie "Her" (Jonze, 2013).

3 Much hope is now pinned on harnessing artificial intelligence to *simplify* and automate legal regulations.

4 here and elsewhere there is an inconsistency in the use of British vs American English spelling, which results from the fact that American English was the spelling used by ChatGPT during the conversation.

References

Aberra, T. (2012). *Michel Foucault power knowledge nexus: Critical analysis and its relevance globalization and current issues of Africa.* LAP LAMBERT Academic Publishing.

Adorno, T. W., & Horkheimer, M. (1944). *Dialectic of enlightenment.* New York, NY: Social Studies Association, Inc.

Agamben, G. (1998). *Homo sacer. Sovereign power and bare life*. Stanford, CA: Stanford University Press.

Agger, B. (2015). *Oversharing: Presentations of self in the Internet age*. Routledge.

Airoldi, M. (2022). *Machine Habitus: Toward a sociology of algorithms*. Cambridge: Polity Press.

Andrejevic, M. (2013). *Infoglut: How too much information is changing the way we think and know*. Routledge.

Atkin, A. (2023). *Peirce's theory of signs*. In E. N. Zalta & U. Nodelman (Eds.), *The Stanford encyclopedia of philosophy* (Spring 2023 ed.).

Austin, J. L. (1975). *How to do things with words* (2nd ed., M. Sbisà & J. O. Urmson, Eds.). Oxford: Oxford University Press.

Bard, A., & Söderqvist, J. (2002). *Netocracy: The new power elite and life after capitalism.* London: Pearson Education.

BBC. (2016). *Norway consumer body stages live app terms reading*. https://www.bbc.co.uk/news/world-europe-36378215

Beer, D. (2009). Power through the algorithm? Participatory web cultures and the technological unconscious. *New Media & Society, 11*(6), 985–1002.

Berger, P., & Luckmann, T. (1971). *The social construction of reality*. Penguin University Books.

Betancourt, M. (2016). *The critique of digital capitalism: An analysis of the political economy of digital culture and technology*. Brooklyn, NY: Punctum Books.

Bilić, P., (2018). A critique of the political economy of algorithms: A brief history of Google's technological rationality. *tripleC: Communication, Capitalism & Critique, 16*(1), 315–331.

Bodenhorn, B., & Vom Bruck, G. (2006). "Entangled in histories": An introduction to the anthropology of names and naming. In G. Vom Bruck & B. Bodenhorn (Eds.), *An anthropology of names and naming* (pp. 1–30). Cambridge: Cambridge University Press. https://doi.org/10.1017/CBO9780511499630.002

Brand, M., Döring, N., Krämer, N. Kruger, T. H. C., Mikhailova, V., & Vowe, G. (2021). Sexual interaction in digital contexts and its implications for sexual health: A conceptual analysis. *Frontiers in Psychology, 12*, 769732. https://doi.org/10.3389/fpsyg.2021.769732

Brooks, D. (2013). The philosophy of data. *The New York Times*. bit.ly/45yGjFy

Broussard, M. (2018). *Artificial unintelligence: How computers misunderstand the World*. Cambridge, MA: MIT Press.

Carr, N. (2014). *The glass cage: How our computers are changing us*. New York, NY: W. W. Norton & Company.

Chaitin, G. J. (1987). *Algorithmic information theory*. Cambridge: Cambridge University Press.

ChatGPT. (2023). Interviewed by R. Zarzycki. OpenAI platform.

Coveney, P., & Highfield, R. (1995). *Frontiers of complexity: The search for order in a chaotic world*. Faber and Faber.

Eubanks, V. (2018). *Automating inequality: How high-tech tools profile. Police, and punish the poor*. New York, NY: St. Martin's Press.

Fagan, P. F. (2009). The effects of pornography on individuals, marriage, family and community. *Research Synthesis*, RS09K01.

Farroni, T., Longa, L. D., & Valori, I. (2021). Interpersonal affective touch in a virtual world: Feeling the social presence of others to overcome loneliness. *Frontiers in Psychology Sec . Human-Media Interaction, 12*, https://doi.org/10.3389/fpsyg.2021.795283

Foucault, M. (1980) Power/knowledge: Selected Interviews and Other Writings 1972–1977. In C. Gordon (Ed.), C. Gordon et al. (Trans.). New York, NY: Pantheon Books.

Foucault, M. (2008). *The history of sexuality*. Penguin.

Fricker, M. (2007). *Epistemic injustice: Power and the ethics of knowing*. Oxford: Oxford University Press. https://doi.org/10.1093/acprof:oso/9780198237907.001.0001

Fridlund, P. (2020). Post-truth politics, performatives and the force. *Jus Cogens, 2*, 215–235. https://doi.org/10.1007/s42439-020-00029-8

Fuller, S. (2018). *Post-truth: Knowledge as a power game*. Anthem Press. https://doi.org/10.2307/j.ctvgd30v

Godler, Y. (2020). Post-post-truth: An adaptationist theory of journalistic verism. *Communication Theory, 30*(2), 169–187. https://doi.org/10.1093/ct/qtz031

Goldfarb, J. (2006). *The politics of small things*. Chicago, IL/London: The University of Chicago Press.

Glynos, J., & Howarth, D., (2007). *Logics of critical explanation in social and political theory*. Routledge.

Gudonis, M. (2017). *How useful is the concept of post-truth in analysing Genocide Denial? Analysis of online comments on the Jedwabne Massacre*. Zoon Politikon 2017/8, 141–182. https://doi.org/10.4467/2543408XZOP.17.006.9265

Hagmann, D., Ho, E., & Loewenstein, G. (2020). Measuring information preferences. *Management Science, 67*(1). https://doi.org/10.1287/mnsc.2019.3543

Jonze, S. (2013). *Her, a movie*. Annapurna Pictures.

Karsten. L. (2020). Counterurbanisation: Why settled families move out of the city again. *Journal of Housing and the Built Environment, 35*(4). https://doi.org/10.1007/s10901-020-09739-3

Kauffman, S. A. (2013). *Evolution beyond Newton, Darwin, and Entailing Law: The origin of complexity in the evolving biosphere.* In C. H. Lineweaver, P. C. W. Davis, & M. Ruse (Eds.), *Complexity and the arrow of time* (pp. 162–190).Cambridge: Cambridge University Press.

Kelly-Smith, R. (2002). Saving lives should pay more than kicking balls. *The Guardian.* bit.ly/43sewES

Keyes, R. (2004). *The post-truth era: Dishonesty and deception in contemporary life.* New York, NY: St. Martin's Press.

Kreft, J. (2018). *Władza algorytmów* (Eng.: *The power of the algorithms*). Cracow: Wydawnictwo Uniwersytetu Jagiellońskiego.

Lasch, C. (2018). *The culture of narcissism.* London/New York, NY: W. W. Norton & Company.

Leichkt, K. T, (2020, December). *Occupations and inequalities in the 21st century: What's in your wallet? Research in Social Stratification and Mobility, 70*, 100550. https://doi.org/10.1016/j.rssm.2020.100550

Lemma, A. (2021). Introduction – Becoming sexual in digital times: The risks and harms of online pornography. *The Psychoanalytic Study of the Child, 74*(1), 118–130. https://doi.org/10.1080/00797308.2020.1859283

Lévi-Strauss, C. (1950). *Introduction to the work of Marcel Mauss* (F. Baker, Trans.). London: Routledge & Kegan Paul.

Lowrey, A. (2019). $350,000 a Year, and just getting by. *The Atlantic.* bit.ly/43NTWPF)

Mager, A. (2012). Algorithmic ideology: How capitalist society shapes search engines. *Information, Communication & Society, 15*(5), 769–787.

Mainzer, M. (1994). *Thinking in complexity.* Berlin: Springer-Verlag.

Maslow, A. H. (1943). A theory of human motivation. *Psychological Review, 50*(4), 370–396. https://doi.org/10.1037/h0054346

McCallum, S. (2023). ChatGPT banned in Italy over privacy concerns. *BBC.* bit.ly/3qfHTfm

McPhee, J. (1989). *The control of nature.* New York, NY: Farrar Straus and Giroux.

Ninno, D., & Totaro, P. (2014). The concept of algorithm as an interpretative key of modern rationality. *Theory, Culture & Society, 31*(4), 29–49.

Penrose, R. (1989). *The emperor's new mind: Concerning computers, minds and the laws of physics.* Oxford University Press.

Piketty, T. (2014). *Capital in the twenty-first century.* Cambridge, MA: The Belknap Press of Harvard University Press.

Pinker, S. (2007). *The stuff of thought. Language as a Window into Human Nature.* Penguin Books.

Rodrigues, R. (2020). Legal and human rights issues of AI: Gaps, challenges and vulnerabilities. *Journal of Responsible Technology, 4*.

Rorty, R. (1989). *Contingency, irony, and solidarity*. Cambridge: Cambridge University Press. https://doi.org/10.1017/CBO9780511804397

Seebach, S. (2017). *Love and society. Special social forms and the master emotion*. London: Routledge. https://doi.org/10.4324/9781315753560

United Nations Conference on Trade and Development (UNCTAD). (2021). *Digital economy report 2021*. New York, NY: United Nations Publications.

Wheeler, K. (2019). Moral economies of consumption. *Journal of Consumer Culture, 19*(2), 271–288.

Whorf, B. L. (2012). Language, Thought, and Reality: Selected Writings of Benjamin Lee Whorf. B. Carroll, S. C. Levinson, & P. Lee (Eds.), The MIT Press.

Wittgenstein, L. (2021). *Tractatus Logico-Philosophicus* (Centenary ed.). Anthem Press. https://doi.org/10.2307/j.ctv22d4t7n

Wolfram, S. (2023). *What is ChatGPT doing … and why does it work?* bit.ly/3qewDzM

Zarzycki, R. (2021, March 12). *Ciasteczkowa farsa* [Cookie farce]. *Polska The Times* (Paper edition).

Zarzycki, R. (2013). *Information inflation and spectacularization of meanings as sources of hyperreality*. In K. Dąbrowska, Ł. Huculak, J. Jernajczyk, B. Skowron, M. Zakrzewska, & R. Zarzycki (Eds.), *Excess and lack* (J. Burzyńska, Trans., pp. 161–171). Wrocław: Wrocław Academic Hub.

Zuboff, S. (2019). *The age of surveillance capitalism: The fight for a human future at the new Frontier of power*. New York, NY: PublicAffairs.

CHAPTER 19

Some Comments Regarding the Digitalisation of the Health Care System in Poland during the COVID-19 State of Epidemic

Selected Legal Aspects

Katarzyna M. Zoń

Abstract

In modern times, the development of science and technological progress have together created new prospects for the provision of health services. These opportunities are of key importance to the functioning of healthcare systems themselves. This article is devoted to analysing the legal dimension of the digitalisation of the healthcare system in Poland, along with the consequences of such a process. The validity and usefulness of the subject matter are justified by the current pandemic situation. In this context, it should be emphasised that the above-mentioned process has recently accelerated. The implementation of this solution is aimed at improving the effectiveness of healthcare service provision. Digitalisation therefore needs to be discussed with reference to its role as an important component of the healthcare system in Poland.

Keywords

health services – telemedicine – teleconsultations – electronic referrals – COVID-19

1 Introduction

Modern technological progress creates new opportunities and perspectives for the implementation of innovations in many areas. The introduced solutions also have a key impact on the functioning of the health care system. A clear indication of the above changes is the digitalisation process, the implementation of which has accelerated significantly in this field recently.

This observable tendency justifies the analysis of selected legal aspects of digitalisation of the public health care system in Poland by taking into consideration how this process occurred during the COVID-19 epidemic.[1] The validity and usefulness of the subject matter are confirmed by the practical significance and topicality of the issues raised. During this unique epidemic situation, the key challenge was to properly organise the form of providing health care services. On the one hand, the availability and continuity of health services had to be ensured, and – on the other – the safety of health professionals and patients had to be guaranteed. The aim of this paper is to draw attention to Polish legal regulations constituting the digitalisation process in the provision of health services, which were introduced during the COVID-19 pandemic. Although modern solutions were identified within the time frame of this specific period, in reality, they are part of the earlier initiated digitalisation process of the health care system in Poland. A comprehensive approach to the considerations also requires indicating the tools introduced earlier and providing an overview of the regulatory environment in existence at the time. The conclusions systematise observations and remarks concerning the review of select legal regulations from the perspective of the current moment.

2 Outline of the State of the Digitalisation of the Health Care System in Poland before the COVID-19 State of the Epidemic

As already mentioned, even before the COVID-19 pandemic, there were solutions in place in Poland that constituted part of the digitalisation process of the public health care system. Taking into account the broader regulatory framework, it is worth mentioning that since 12th December 2015, the Polish legislator has widely allowed the possibility of providing health care services in the telemedicine model, i.e., via ICT systems or communication systems.[2] Necessary legal provisions have been provided for in the Medical Care Act,[3] among others, and in acts relating to the exercising of individual medical professions.[4] However, issues related to its financing still remain an important practical barrier to the widespread provision of telemedicine in the public healthcare system. Despite the introduction of the general possibility of providing such health services, only some of them were covered by public financing. This applies, for example, to cardiac teleconsultation, geriatric teleconsultation, hybrid cardiac telerehabilitation and medical consultation in primary health care carried out remotely, i.e., via ICT systems or communication systems.[5] The latter has gained particular importance during the COVID-19 pandemic era, which will be presented below as part of this paper. The aforementioned general provisions were linked to specific legislative changes in the area of health care digitalisation aimed at improving the health care services provision process. The key in this respect was the successive implementation of subsequent process phases, i.e., medical sick e-notes[6] and e-prescriptions.[7] An Internet Patient Account (Pol.: Internetowe Konto Pacjenta, IKP) was also launched,[8] with the scope of its functionality extended over time. This tool allows patients to access their medical records, information about issued sick leaves, completed referrals, prescriptions, as well as about their current entitlement to healthcare services or the amount of health insurance premiums paid by the patient. Moreover, via IKP, the patient can authorise another person to access his or her medical data or information regarding health, as well as give consent to the provision of health services.

3 Select Examples of Digitalisation of the Polish Health Care System during the COVID-19 State of the Epidemic

3.1 *Electronic Referrals* (*E-Referrals*)

With the aforementioned in mind, it is worth expanding on the outlined system of legal regulations with norms constituting aspects of the health care system digitalisation process in Poland, and which came into force during the COVID-19 pandemic. It should be indicated that these are not of a uniform nature. The first group includes solutions introduced at a predetermined time, which were introduced irrespective of the specific epidemic situation. Conversely, the second category comprises instruments implemented in connection with the provision of health care services in these specific circumstances.

With regards to e-referrals,[9] which is an aspect of the health care system digitalisation that falls within the first of the above groups, it should be clarified that the obligation, existing since 8th January 2021, to issue referrals in electronic form applies only to specific documents, recorded in the Medical Information System (Pol.: System Informacji Medycznej SIM). This is because not all referrals issued to enable the patient to access specific health services are of such a nature. The list established by way of a regulation of the Minister of Health[10] lists specific types of medical services, taking into account the manner of their financing and the rules of their implementation. The electronic form referral includes: outpatient specialist services financed from public funds, excluding specialist advice – speech therapy, as well as publicly funded fetal echocardiography examinations, publicly funded gastrointestinal endoscopic examinations, publicly and non-publicly funded nuclear medicine examinations, publicly funded magnetic resonance imaging examinations, publicly and non-publicly funded computed tomography examinations, hospital treatment, therapeutic rehabilitation, health resort treatment and rehabilitation. However, referrals for other services, such as drug programmes, can still be issued in paper form. Similarly, this encompasses diagnostic tests or certain procedures that are performed for a patient as part of specialist advice, hospitalisation, as well as in the absence of access to the ICT system (for example, system failure) and in the case of issuing a referral for a person of unknown identity.

3.2 *Electronic Medical Records*

Another indication of digitalisation, not directly resulting from the unique epidemiological situation, can be seen in the field of electronic medical records. However, the aforementioned solution is not a brand-new instrument. As early as 1st January 2019, healthcare institutions were obliged to keep certain documents in an electronic form such as disease, health problem or injury diagnosis information, test results, refusal of hospital admission justifications, health services provided and possible medical orders – in the case of a patient being refused admission to hospital, information for the doctor referring the recipient to a specialist clinic or hospital treatment concerning the diagnosis, treatment method, prognosis, prescribed drugs, foodstuffs for special nutritional purposes and medical devices, including the usage period and dosage method, as well as scheduled control visits and information sheets from hospital treatment.[11]

Over time, the above obligation incorporated additional documents, e.g., laboratory test results with descriptions and diagnostic test descriptions other than the above. Electronic medical records[12] are documents created in an electronic form with a qualified electronic signature, a trusted e-signature, a personal signature or by way of confirming the origin and integrity of data available in the ICT system made available free of charge by the Social Insurance Institution, i.e., prescriptions, referrals, orders for supplies and repairs and vaccination cards, as well as documents which it was obligatory to keep in the form of electronic records at an earlier date.[13] In addition, since 1st July 2021, the healthcare provider is obliged to ensure that it is possible to exchange the data contained in the electronic medical records via the SIM system.[14] The listed catalogue of documents is to be expanded to include all documentation maintained by the healthcare provider. The above solution is intended to improve the system of medical record circulation between healthcare providers, guaranteeing proper cooperation and giving assurance of comprehensive care for patients. Automatic access to electronic medical records is available to designated entities: the medical employee who created the electronic medical records containing the patient's personal data or individual medical data, the medical employee practising at the service provider where electronic medical records containing a patient's personal data or individual medical data has been created in connection with his/her practising at this service provider if it is necessary to carry out diagnostics or ensure continuity of treatment, as well as the doctor, nurse or midwife providing healthcare services to the service recipient under the healthcare services provision agreement in the field of primary healthcare and all medical personnel in a situation that is life-threatening for the patient.[15] In other cases, access to medical records requires the consent of the patient or his or her legal representative, while indicating the scope of time and subject matter of data access.

3.3 *Teleconsultations*

Given the need to reduce direct contact between patients and medical professionals in the COVID-19 pandemic era, teleconsultation provided in primary care has become essential. Crucially, these services have been included as part of publicly funded benefits since November 2019. A teleconsultation is a health service provided remotely using ICT or communication systems, and carried out as part of primary health care (PHC) by a doctor, nurse or midwife who provides services at a PHC provider. This can especially take the form of a conversation over the phone, a video conference or exchange of electronic correspondence (emails). The dynamic increase in the amounts of teleconsultations practiced observed during the COVID-19 epidemic directly influenced the necessity to develop organisational standards regarding their provision. Within the scope of the implemented principles, which changed three times during the time the state of the epidemic was in force,[16] the duties of the primary care provider and the teleconsultation provider have been listed. These include, above all, obligations towards the patient regarding the provision of information. One such obligation that should be emphasised is providing the patient with information about the terms and conditions of teleconsultation provision, including the patient's right to express their wish to have personal contact with relevant medical staff, specifying, e.g., the services to be provided through direct patient contact only, how the date and time of the teleconsultation is to be determined or how to establish contact with the patient. It is also worth mentioning their report duties towards the National Health Fund, which in turn allow the latter to monitor the health services provided in this form, but also it should be indicated that there is a necessity to record the provision of teleconsultation services in the medical documentation and to conduct it maintaining confidentiality. Wider use of teleconsultation in the pandemic was also fostered by complementary digitalisation tools, such as the aforementioned e-prescriptions, e-referrals or e-notes.

3.4 *Public Financing of Health Services Provided under the Telemedicine Model*

The COVID-19 pandemic also increased the scope of publicly funded health services delivered using ICT or communications systems. Some of them were reimbursements covered regardless of the epidemic situation, while others only during the state of epidemiological threat or the state of epidemic emergency. The first group includes such services as the provision of advice and visits in the field of palliative and hospice care in a home setting with the use of ICT systems or other communication systems, provided that such a procedure does not pose a risk of causing the deterioration of the patient's health condition,[17] guaranteed benefits within the scope of nursing and home care services within the framework of long-term care, provided under home conditions by a long-term home care team or long-term home care nursing service,[18] as well as outpatient guaranteed benefits within the scope of psychiatric care and community (home) mental treatment and addiction treatment, which can be provided with the use of ICT systems on condition that the availability of personnel required to provide them at the place of benefit provision is ensured.[19] Other services have only been approved temporarily (during the state of epidemiological threat or the state of epidemic emergency). They include the provision of guaranteed benefits (within the scope of haemodialysis and haemodialysis with 24-hour stand-by services) by an authorised physician, using the ICT systems or communication systems, if this manner of provision of benefits does not pose a threat to a patient's life or could lead to a deterioration of his/her health condition, and a nurse with appropriate qualifications is present at the patient's side.[20] This also covers the scope of guaranteed benefits in the area of therapeutic rehabilitation, provided under conditions of outpatient, home or day wards, a possibility of providing medical advice, with counselling or therapeutic visits using ICT systems or communication systems having been admitted, if this method of providing healthcare does not pose a threat to the deterioration of the health condition of the benefit recipient.[21] Due to the cessation of the special circumstances (state of an epidemic or an epidemic threat), there are currently no legal bases for the refunding of the above services provided with the use of information and communication (ICT) systems.

4 Patient Identification and Confirmation of Entitlement to Benefits

During the COVID-19 pandemic, there were several additional solutions provided for facilitating the identification of a patient or confirming eligibility for a given service,[22] the introduction of which was directly related to the existing epidemiological situation. On this basis, entities providing healthcare services via ICT systems or communications systems were allowed to confirm the identity of the recipient based on data provided by the recipient via those systems, including by telephone. In addition, certain formal obligations concerning the provision of health care services did not apply to appeals (among others, the obligation to submit to the provider within 14 days the original copy of the referral as part of health care services financed from public funds). It was also made possible to issue instructions for the provision of medical devices and repair orders via ICT systems or communication systems. The functionality of the above-mentioned tools, which in some situations may also have proven useful after the end of the extraordinary epidemiological situation justifying their implementation, has not been maintained.[23]

5 Conclusions

With these considerations in mind, it should be emphasized that digitalisation continues to constitute a significant challenge for the public health care system in Poland. The accomplishment of subsequent stages of this process is aimed at streamlining and improving the efficiency of healthcare services, enabling the implementation of personalized patient care. Moreover, progressing digitalisation fosters the further development of telemedicine, for which it remains extremely important to solve the issue of the financing of this type of health service provision.

The announcement of the COVID-19 state of epidemic made it necessary to reach for other ways of providing health care services, serving as an impetus for a more in-depth analysis of the functioning of the health care system in Poland. For epidemiological reasons, ensuring access to health care had to be accompanied by simultaneous efforts to minimise the risk of infection among medical personnel and patients. The experience gained during the epidemic, related to the need to reduce direct contacts has highlighted the pivotal role played by digitalisation in the overall functionality of the health care system. There is no doubt that it is an inevitable process for the creation of a modern model of the health care system serving the needs of both patients and medical professionals. In this context, it is worth remembering that technology should strengthen the traditional health service model and not seek to replace it. The COVID-19 state of epidemic undoubtedly contributed to wider use of previously functioning tools, whose potential had not been recognised before. A manifestation of this was, for example, the dynamic increase in the use of teleconsultation in the diagnostic and therapeutic process. It remains an open matter to what extent the solutions developed in the area of health care will function after the extraordinary epidemic circumstances have ceased to exist. However, an assessment of this aspect requires further consideration in the future, after some time has passed.

Notes

1 The state of epidemic in Poland lasted from 20 March 2020 to 15 May 2022.

2 Act of 9th October 2015 amending the information system in health care act and some other acts (Journal of Laws, item 1991).

3 Cf. Art. 3 (1) of the Medical Care Act of 15th April 2011 (consolidated text Journal of Laws 2023, item 911 as amended).

4 Cf. Article 2(4) of the Act of 5th December 1996 on the doctor and dentist professions (consolidated text Journal of Laws 2023, item 1516 as amended), Article 11(1) of the Act of 15th July 2011 on the nurse and midwife professions (consolidated text Journal of Laws 2022, item 2702).

5 Regulation of the Minister of Health of 31st October 2019 amending the regulation on guaranteed services in the field of primary health care (Journal of Laws, item 2120).

6 Cf. Article 55 of the Act of 25th June 1999 on cash benefits from social insurance in the event of sickness and maternity (consolidated text Journal of Laws 2022, item 1732). Sick leave in the form of an electronic document (e-note) has been issued since 1st January 2016 (parallel with a paper form). However, as of 1st December 2018, this e-note is now the exclusive form.

7 Cf. Article 95b of the Act of 6th September 2001. Pharmaceutical Law (consolidated text Journal of Laws 2022 item 2301 as amended). Prescriptions in electronic form (e-prescriptions) have been issued since 18th April 2018. This form has become obligatory since 1st January 2020 with some exceptions, e.g., the possibility to issue paper prescriptions e.g., for persons of unknown identity or *pro auctore* prescriptions (for the prescribing doctor himself) and *pro familiae* prescriptions (for persons close to the prescribing doctor – spouses, ascendants, descendants or siblings).

8 The Internet Patient Account has been functioning in the Polish legal framework since 23rd August 2018 (Act of 20th July 2018 on amending the information system in health care act and some other acts, Journal of Laws 2018, item 1515).

9 Cf. Article 59aa of the Act of 27th August 2004 on health care services financed from public funds (consolidated text Journal of Laws 2022, item 2561 as amended). Referrals in electronic form (e-referrals) have been issued since 1st April 2019. This form has become the exclusive form since 1st January 2021 with regards to documents recorded in the Medical Information System.

10 Cf. § 2 of the Regulation of the Minister of Health of 15th April 2019 on referrals issued in electronic form in the Medical Information System (consolidated text Journal of Laws 2022, item 1417 as amended), which came into force on 18th April 2019.

11 Cf. Article 56(1) of the Act of 28th April 2011 on the information system in health care (consolidated text Journal of Laws 2022, item 1555 as amended).

12 Cf. Article 2(6) of the Act of 28th April 2011 on the information system in health care (consolidated text Journal of Laws 2022, item 1555 as amended).

13 Cf. § 1 Regulation of the Minister of Health of 8th May 2018 on types of electronic medical records (consolidated text Journal of Laws 2023, item 1851), which entered into force on 2nd June 2018.
14 Cf. Article 56(4) of the Act of 28th April 2011 on the information system in health care (consolidated text Journal of Laws 2022, item 1555 as amended).
15 Cf. Article 35 (1) and (1a) of the Act of 28th April 2011 on the information system in health care (consolidated text Journal of Laws 2022, item 1555 as amended).
16 Regulation of The Minister of Health of 12th August 2020 on the organisational standard of teleconsultation in primary health care (consolidated text Journal of Laws 2022, item 1194).
17 Regulation of the Minister of Health of 16th March 2020 amending regulation on the guaranteed services in the field of palliative and hospice care (Journal of Laws, item 457), which entered into force on 17th March 2020.
18 Regulation of the Minister of Health of 16th March 2020 amending the regulation on guaranteed benefits in the field of nursing and care services in long-term care (Journal of Laws 2020, item 460), which entered into force on 17th March 2020.
19 Regulation of the Minister of Health of 16th March 2020 amending the regulation on guaranteed benefits in the field of psychiatric care and addiction treatment (Journal of Laws, item 456), which entered into force on 17th March 2020.
20 Regulation of the Minister of Health of 6th April 2020 amending the regulation on guaranteed benefits in the field of outpatient specialist care (Journal of Laws, item 612), which entered into force on 8th April 2020.
21 Regulation of the Minister of Health of 10th April 2020 amending the regulation on guaranteed services in the field of medical rehabilitation (Journal of Laws, item 660), which entered into force on 11th April 2020.
22 The legal basis for their functioning was the Regulation of the Council of Ministers on establishing certain restrictions, orders and prohibitions in connection with the occurrence of an epidemic.
23 Regulation of the Council of Ministers of 27 June 2023 repealing the Regulation on the establishment of certain restrictions, orders and prohibitions in connection with the occurrence of an epidemic emergency (Journal of Laws, item 1233), which entered into force on 1 July 2023.

References

Act of 5th December 1996 on the doctor and dentist professions (consolidated text Journal of Laws 2023, item 1516 as amended).

Act of 25th June 1999 on cash benefits from social insurance in the event of sickness and maternity (consolidated text Journal of Laws 2022, item 1732 as amended).

Act of 6th September 2001 Pharmaceutical Law (consolidated text Journal of Laws 2022 item 2301 as amended).

Act of 27th August 2004 on health care services financed from public funds (consolidated text Journal of Laws 2022, item 2561 as amended).

Medical Care Act of 15th April 2011 (consolidated text Journal of Laws 2023, item 911 as amended).

Act of 28th April 2011 on the information system in health care (consolidated text Journal of Laws 2022, item 1555 as amended).

Act of 15th July 2011 on the nurse and midwife professions (consolidated text Journal of Laws 2022, item 2702 as amended).

Act of 9th October 2015 amending the information system in health care act and some other acts (Journal of Laws, item 1991).

Act of 20th July 2018 on amending the information system in health care act and some other acts (Journal of Laws, 2018 item 1515).

Regulation of the Council of Ministers of 27th June 2023 repealing the Regulation on the establishment of certain restrictions, orders and prohibitions in connection with the occurrence of an epidemic emergency (Journal of Laws, item 1233).

Regulation of the Minister of Health of 8th May 2018 on types of electronic medical records (consolidated text Journal of Laws 2023, item 185).

Regulation of the Minister of Health of 15th April 2019 on referrals issued in electronic form in the Medical Information System (consolidated text Journal of Laws 2022, item 1417 as amended).

Regulation of the Minister of Health of 31st October 2019 amending the guaranteed services in the field of primary health care (Journal of Laws, item 2120).

Regulation of the Minister of Health of 16th March 2020 amending the guaranteed benefits in the field of psychiatric care and addiction treatment decree (Journal of Laws, item 456).

Regulation of the Minister of Health of 16th March 2020 amending the guaranteed services in the field of palliative and hospice care decree (Journal of Laws, item 457).

Regulation of the Minister of Health of 16th March 2020 amending the guaranteed benefits in the field of nursing and care services in long-term care decree (Journal of Laws, item 460).

Regulation of the Minister of Health of 6th April 2020 amending the guaranteed benefits in the field of outpatient specialist care decree (Journal of Laws, item 612).

Regulation of the Minister of Health of 10th April 2020 amending the guaranteed services in the field of medical rehabilitation decree (Journal of Laws, item 660).

Regulation of the Minister of Health of 12th August 2020 on the organisational standard of teleconsultation in primary health care (consolidated text Journal of Laws 2022, item 1194).

CHAPTER 20

Works of Art

Contributing Artists

ART 1 Antonina Joszczuk-Brzozowska, *Materialization*, 2020 (hand-formed hot glass, 40 × 70 × 20 cm)[a]
PHOTOGRAPH: DZIKIE STUDIO

a Work first published in the catalog of the exhibition "GLASS_ON. Let's meet in better Times", Wrocław 2021, pp. 14–15.

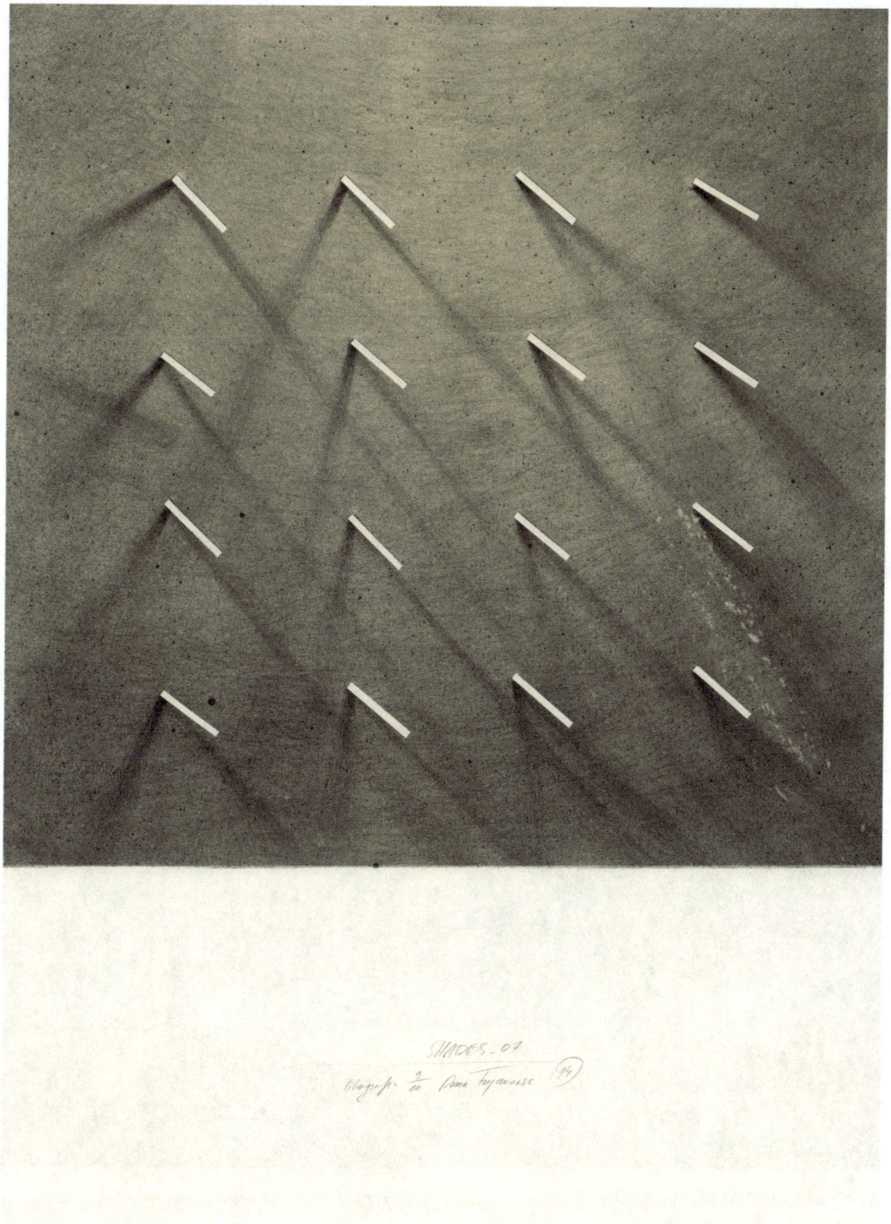

ART 2 Anna Trojanowska, *SHADES_07*, 2014 (lithograph on Carrara marble, 50 × 35 cm)

ART 3 Kamil Moskowczenko, *Third Diagonal Symmetry*, 2020 (oil on canvas, 160 × 120 cm)

ART 4 Beata Mak-Sobota, *Onestring*, 2017 (soda glass float, cut, ground, polished, steel, 35 × 45 × 16 cm)[a]
PHOTOGRAPH: JUSTYNA ŻAK

a Work first published in: "S+C Szkło i Ceramika" nr 3/2021, p. 9.

ART 5 Karolina Freino, *Cataract*, 2016 (intervention in the public space of Katowice)[a]

a Work first published in: Phantom/Widmo, Ed. M. Lisok, Katowice 2016, p. 77.

ART 6 Łukasz Huculak, *Cordyceps*, 2022 (acrilic on canvas, 160 × 120 cm)

ART 7 Zuzanna Dyrda, *I'm Just a Pawn*, 2021 (printmaking performance)[a]

a Work first published in: Urgelles, C. (2023). Printmaking resilience in pandemic times. *IMPACT Printmaking Journal*, no. 6.

ART 8 Dominika Sobolewska, *RGB*, 2008
A spatial arrangement of nine cubes made of transparent plastic in the basic additive colors (area about 12 sqm). Project realized in cooperation with WRO Art Center.[a]

a Work first published in the catalog of the exhibition "Update_3 | Body sound", Antwerp 2010.

ART 9 Karolina Szymanowska, *Sensitive Point*, 2014 (object)

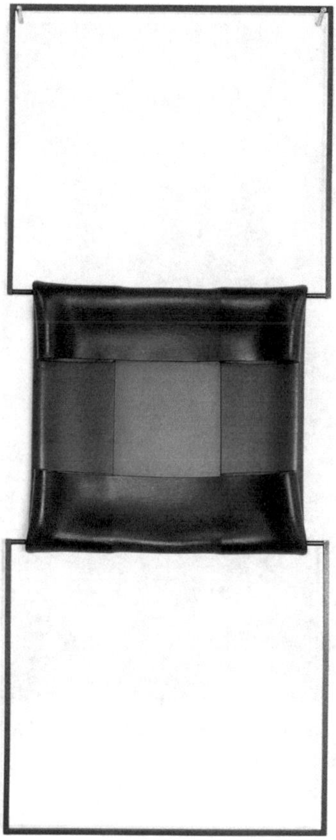

ART 10
Marzena Krzemińska-Baluch, *The Stijl*, 2018 (slumped glass, metal, 96 × 36 × 5 cm)
PHOTOGRAPH: ALICJA KIELAN

ART 11 Bożena Sacharczuk, *Proximity I, II*, from the series *Between...*, 2022 (stoneware molded on a potter's wheel, author's glazes, 65 × 44 and 66 × 43 cm)

ART 12 Anna Kołodziejczyk, *The Power of a Seemingly Simple Interior*, 2016 (acrilic on canvas, 160 × 110 cm)
PHOTOGRAPH: MAŁGORZATA KUJDA

ART 13 Magda Grzybowska, *Sketches for space*, 2015 (installation, steel, textile rubber, 200 × 200 × 100 cm)

ART 14 Mariusz Gorzelak, *Wonder emptiness – I'm waiting for you*, 2020 (intaglio, relief printing, 50 × 70 cm)

ART 15 Ivan Juarez, *Insect City*, Linz, Austria, 2020 (series of wooden elements)[a]
PHOTOGRAPH: GREGOR GRAF

a Work first published in: Juarez, I. (2021). Understanding nature in contemporary art and design. In D. Sobolewska (Ed.), *Signs of empathy in contemporary design* (p. 25). Wrocław.

ART 16 Paweł Baśnik, *Priestess*, 2023 (oil on canvas, 250 × 175 cm)

ART 17 Piotr Jędrzejewski, *67/68*, 2016 (CNC-milled HPL, laser cut steel, 190 × 270 × 12 cm) Two ticking clocks, each at a slightly different pace. For every 67 swings of one pendulum, there are 68 swings of the other. Perhaps time does not always flow at the same speed.[a]
PHOTOGRAPH: JANUSZ BOCHAJCZUK

a Work first published in: Jędrzejewski, P. (2018 Winter). *Ein treffpunkt der Kunst, dem Design und der Uhrmacherei.* "Chronometrophilia", Nr 81 , La Chaux-de-Fonds.

ART 18 Artur Skowroński, *Wrocław Sienkiewicza*, 2017 (digital print, 70 × 50 cm)

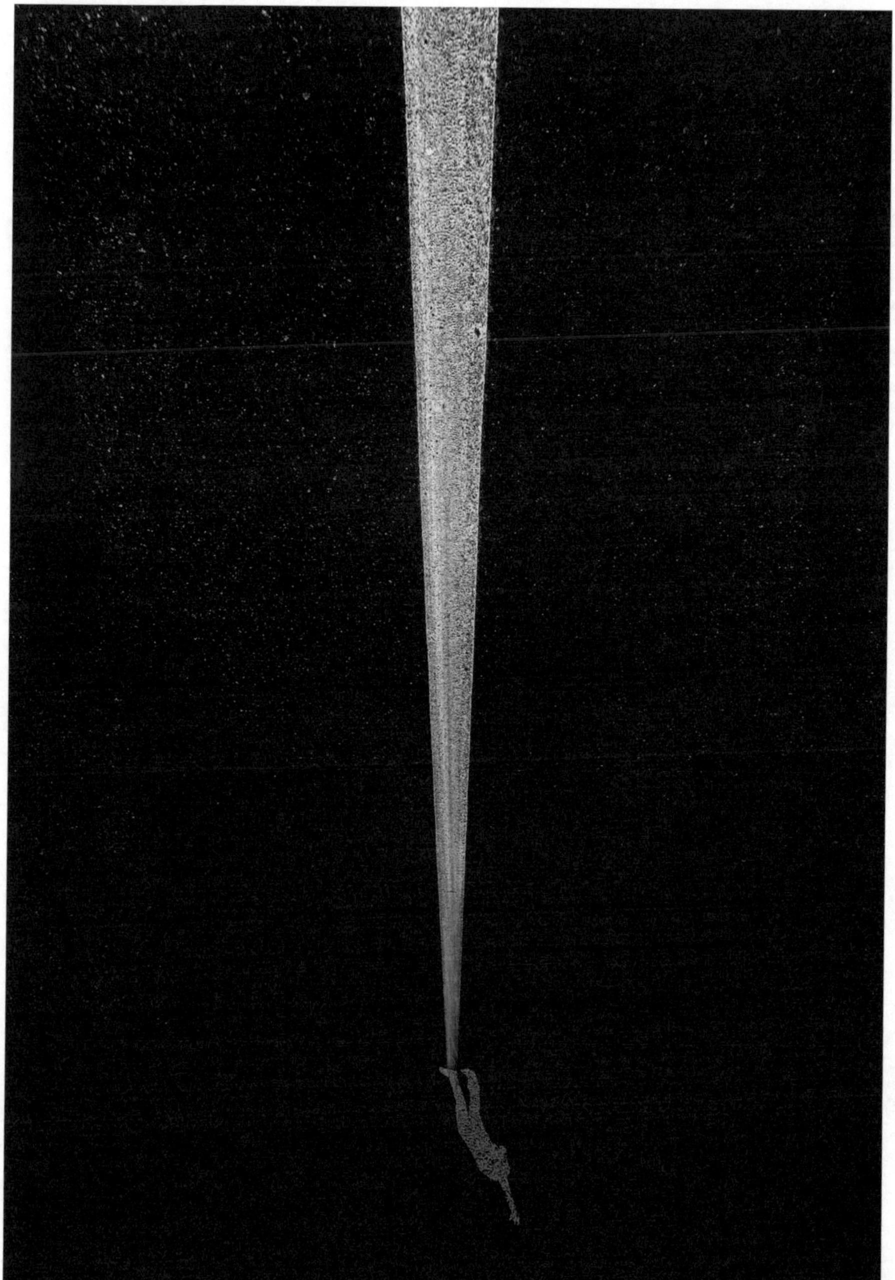

ART 19 Artur Skowroński, *Wrocław Podwale*, 2017 (digital print, 70 × 50 cm)

ART 20 Karina Marusińska, *Cacophony*, from the series *Communication diseases*, 2021 (ceramics, acrylic, resin, diameter 45 cm)[a]
PHOTOGRAPH: GRZEGORZ STADNIK

a Work first published in: Braun, M., & Marusińska, K. (2024). *Okruchy świata. Wobec nieruchomości rzeczy*. In B. Bartecka (Ed.), *O pracach Kariny Marusińskiej* (p. 104). Wrocław.

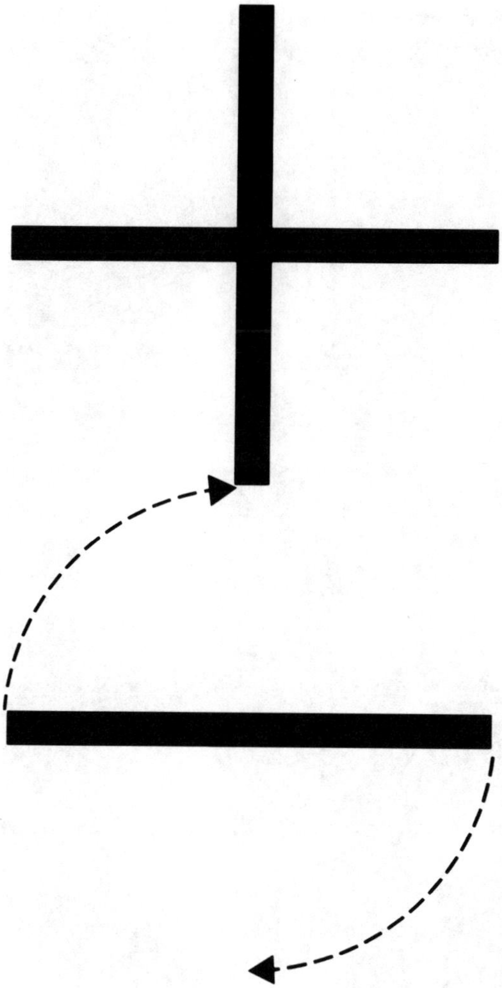

ART 21
Wiesław Gołuch, *plus minus*, 1981/2019
(digital print, 100 × 70 cm)

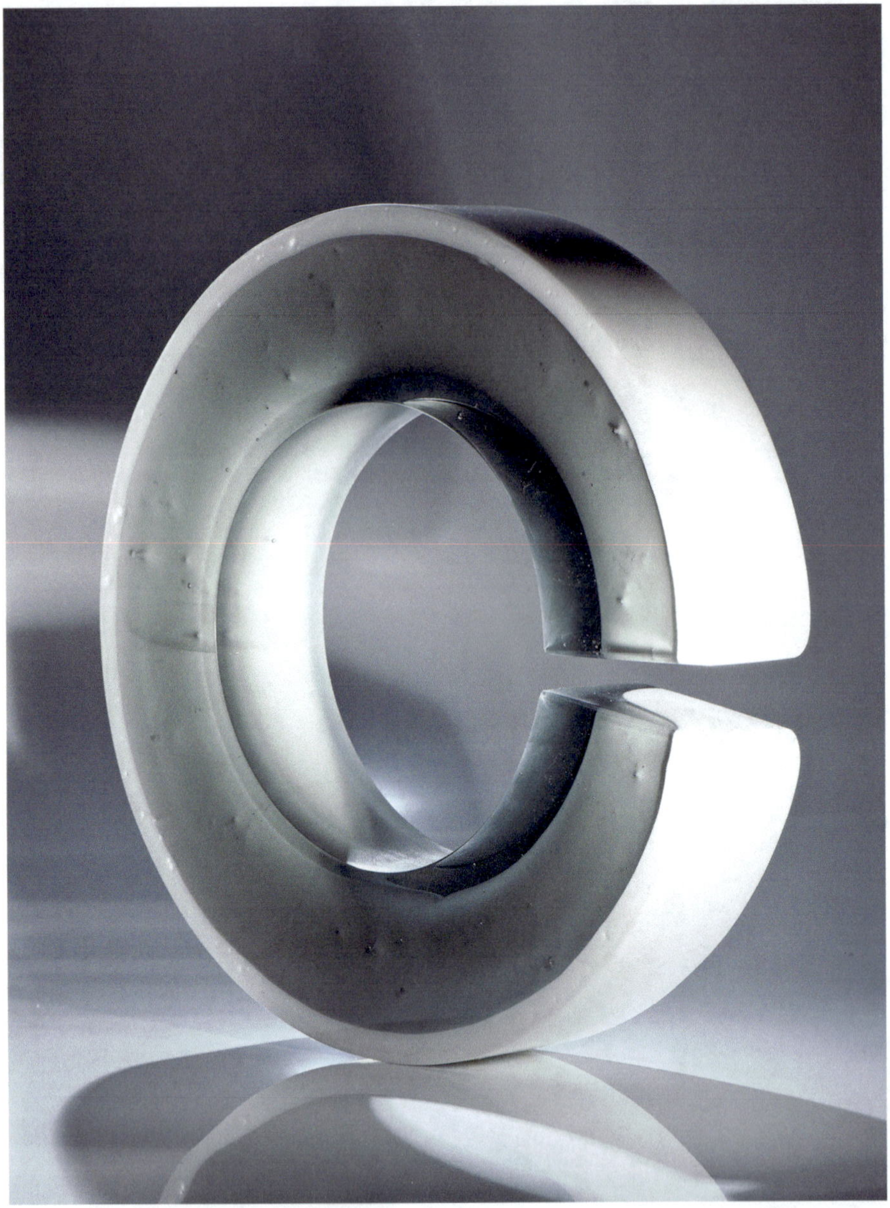

ART 22 Agnieszka Leśniak-Banasiak, *Amoria*, 2022 (kiln-formed glass, coldworked, 41 × 8 cm)[a]
PHOTOGRAPH: KRZYSZTOF PACHURKA

a Work first published in the catalog of the exhibition: "All Quiet on The Western Front? Faculty of Painting and the Department of Glass The Eugeniusz Geppert Academy of Art and Design in Wrocław", Wrocław 2023, p. 68.

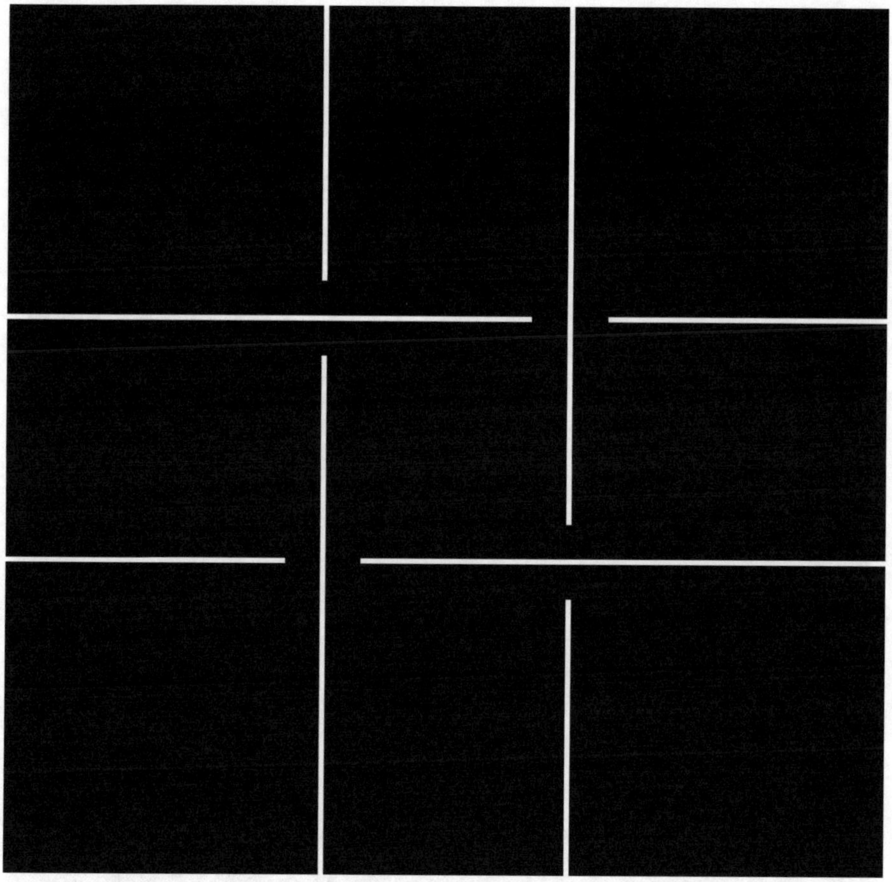

ART 23 Jakub Jernajczyk, *The Knot of Cuts*, 2023 (digital print, 100 × 100 cm)
A dimensionless point can be indicated by cutting a straight line at a certain place.

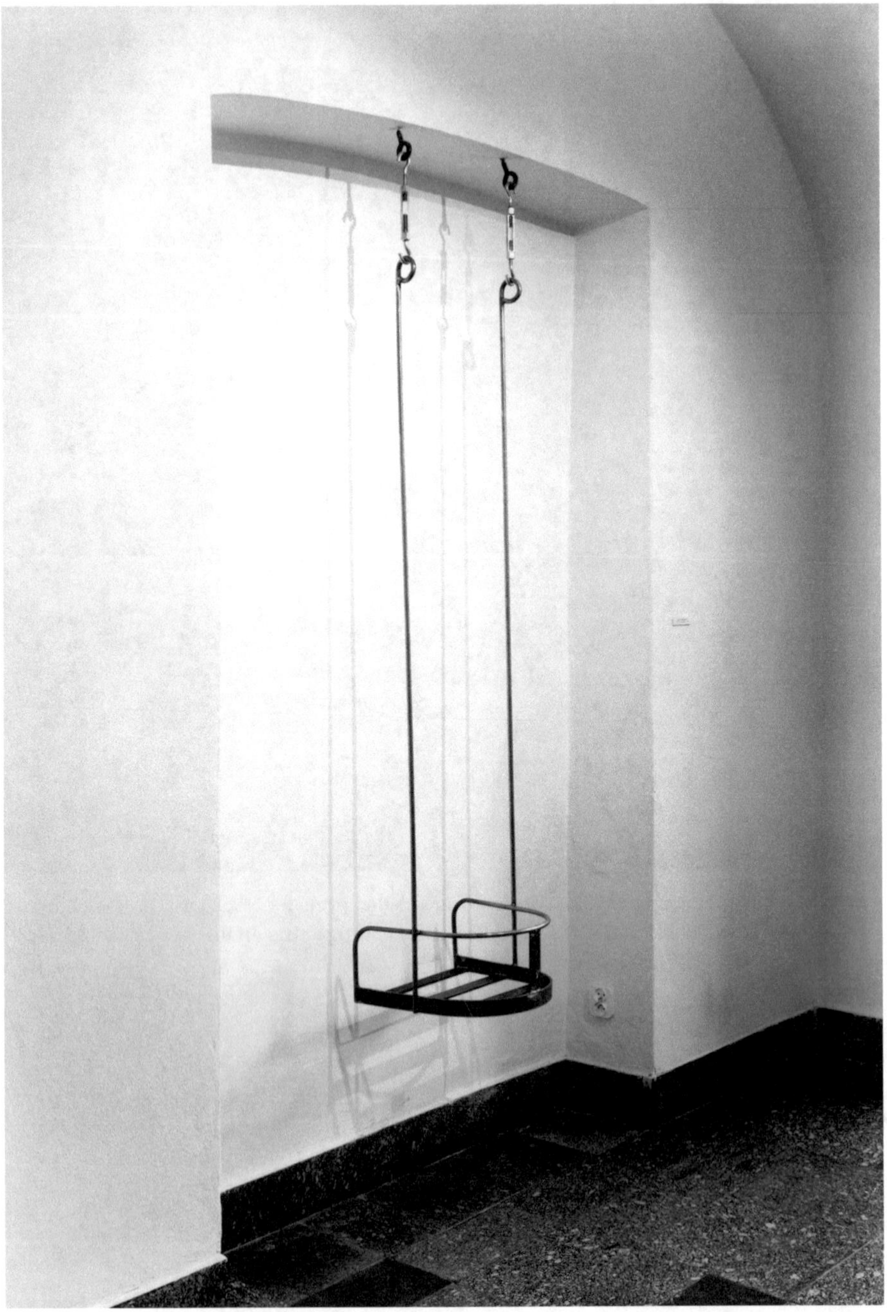

ART 24 Marcin Mierzicki, *PLAY MY BOY*, 2001 (object, steel, 185 × 40 × 45 cm)

ART 25 Mirosław Kociński, *Shield 7*, 2023 (ceramic mosaic [porcelain, stoneware], diameter 58.5 cm)[a]

a Work first published in the catalog of the exhibition: "Qunszt. Ceramika i szkło", Wrocław 2023, p. 28.

ART 26 Eugeniusz Smoliński, *On the matter of simplicity and complexity*, 2014 (digital print, pencil, paper, 21 × 30 cm)
Complexity may be simple. Direct translation from the Polish language: Folded may be straight.[a]

a Work first published in: Miłkowska, D. (2020). Eugeniusz Smoliński, Wrocław, p. 55.

może być
PROST A
Eugeniusz Smolinski

ART 27 Marek Grzyb, *Stories*, 2022 (installation, 80 × 120 × 10 cm)

ART 28 Tomasz Dobiszewski, *Tracking the Sun*, 2017 (laser engraving in marble, 12 slabs on wooden boxes, 30 × 91 × 2 cm each)
The record of the presence of the Sun in a certain latitude, day after day, in each month of the year 2016. Traces burned by the Sun on paper strips were transferred and laser-engraved on twelve marble slabs.

ART 29 Inside Job (Ula Lucińska, Michał Knychaus), *Diaphane*, 2022 (stainless steel, brass, 67 × 45 × 5 cm)

ART 30 Joanna Opalska-Brzecka, *Excess*, 2020 (hand-formed chamotte clay, engobes, 55 × 35 × 50 cm)

ART 31 Piotr Kmita, *The Great Gatsby*, 2013 (digital print, 104 × 385 cm)

ART 32 Justyna Baśnik, *Adoratio*, 2023 (acrilic on canvas, 250 × 180 cm)

ART 33 Justyna Żak, *Written by shadow*, 2014 (Pâte de verre, 29 × 21 cm)

ART 34 Jarosław Grulkowski, *The Expansion*, 2020 (drawing, 100 × 70 cm)

CHAPTER 21

Conclusions

Ewa Błaszczak, Małgorzata Cebrat, Jakub Jernajczyk, Bartłomiej Skowron, Weronika Urbańska and Roland Zarzycki

Our title – *Complexity and Simplicity* – may suggest two distinct methodological approaches: (1) a broad overview of phenomena characterized by either simplicity or complexity, or (2) an analysis of the concepts of complexity and simplicity and the relationships between them. As indicated in the Introduction, we have tried to ensure that the content of this volume meets both of these expectations. On the one hand, we have presented texts describing the significance of simple and complex structures and processes in many different domains of human cognition. On the other, we have sought to point out in clear terms the ontological and epistemological tension between the notions of complexity and simplicity.

1 Tools of Description

From a practical point of view, the concepts of complexity and simplicity can only be regarded as useful linguistic tools. In the hands of scholars and artists, they are the equivalent of what a ladder or a pair of pliers are for a craftsman. Hence, complexity and simplicity also often appear in this volume as operational, functional terms with no clear ontology behind them. Without these concepts, we would not be able to describe a diverse range of phenomena and structures, whether in the natural sciences or the humanities. These terms therefore turn out to be indispensable for organizing and categorizing knowledge in specific disciplines of science and art.

The undeniable tendency to arbitrarily classify phenomena as either simple or complex is largely due to our cognitive limitations. We experience only a very limited slice of reality, and within that narrow spectrum, we make seemingly obvious distinctions between the simple and the complex. We then project these categories, often unconsciously and automatically, onto other parts of reality. Thus, these concepts are not just transparent tools of description, but rather become necessary means of co-creating reality.

2 The Emergence of Meaning

The concepts of complexity and simplicity allow for the relatively easy, though not always meaningful, superposition of structures that previously had little in common. As an exercise, the reader may choose random occurrences of complexity and superimpose them on top of each other in a search for meaning. For example, let us consider ideas taken from the present volume: the complexity of harmonic relations in music, the complexity of a straight line, the complexity of new media, the complexity of philosophical discourse, the complexity of information, the complexity of emotional experience, the structural complexity of amino acids, or the complexity of the movement of the Sun. Their very juxtaposition stimulates the imagination and drives the archetypal content of complexity. Moreover, thanks to such contents, we can make – more or less sensible – mappings of patterns from one conceptual space to another.

The strength of the *Complexity and Simplicity* project lies precisely in the fact that, even before any planned scientific investigation, we can juxtapose the complexity of emotional experience with the structural complexity of amino acids, if only as an exercise. We can, for a moment, impose the order of emotional experience on the order of amino acids and consider whether anything meaningful emerges. Perhaps the adaptation of the structure of emotions to the structure of amino acids, or vice versa, will lead to a creative discovery, the emergence of new meanings or perhaps it will serve as a basis for artistic creation. Of course, it can also become a source of erroneous speculation and inference, of which both scholars and conscious creators, supported by phronetic critical thinking, should always be wary.

3 Levels of Cognition

What appears simple at one level may prove complex when seen from another point of view, another perspective. Nicholas of Cusa teaches that "if it were possible for someone to be situated outside of the world, the world [as such] would be invisible to him, after the fashion of an indivisible point" (Nicholas of Cusa, 2000, p. 10). Practical wisdom, what Aristotle (2011) called *phronesis*, makes us sensitive to the totality of the situation and allows us to act as we should. At one turn, we consider seemingly indivisible and undifferentiated points, and at another we decompose those same indivisible points into infinite structures, creating iterative contexts. In this respect, cognition and action resemble a fractal structure, whose fragments – seemingly simple at a given level – reveal, with each successive enlargement, both new and already known yet potentially and actually infinite, complexity. The graphic representation of Cantor's set provided at the end of this Conclusion may serve as an illustration here. In this most minimalistic fractal, we always see an infinitely complex arrangement of the simplest objects, no matter what level of depth we are at. Again, in line with Aristotle's thought, moving through these levels as one should, and not just as chance suggests, is a matter of the highest art: i.e., practical wisdom.

Complexity and simplicity, when treated solely as tools of description, may appear to be transparent entities, seemingly devoid of any effect on the reality they seek to describe. However, this is not the case. As Hutchins (2012) points out, concepts, while part of rich cognitive ecosystems, arise within a certain practice and die with that practice, just as languages or other systems for describing reality themselves perish. Concepts are part of the distributed cognition that groups of thinkers co-create. They can reinforce or weaken the intellectual mood and, in effect, function to co-create thought collectives (Fleck, 1979) – i.e., certain patterns of directed perception. In this volume, what we propose is to practice thinking and acting, in both science and art, through the lens and with the assistance of the concepts of simplicity and complexity. Why is this important, and what purpose can it serve? Hutchins (2012) has pointed to several mechanisms that increase conceptual order and thus, in a cognitive spiral of order-building and order-destruction, elevate us to a higher conceptual level. This elevation is the essence of this volume. Among these conceptual meta-tools are the aforementioned superposition of structure and mapping across conceptual spaces. In addition, we will indicate other mechanisms mentioned by Hutchins (2012), such as dimensionality reduction, filtering, and composite mechanisms of modulated positive feedback.

Simplicity and complexity make it possible to move between dimensions of cognition: on the one hand, they make it possible to reduce a dimension, and on the other, to deepen it – as in the case of successive layers in a fractal structure. In the context of his philosophy of space, Simmel would say that to move between the dimensions of cognition is, in fact, to approach one dimension and to distance oneself from another (cf. Milà, 2016). Distance and proximity not only constitute the social structure, but also allow us to dynamically approach and separate ourselves from others and, more generally, to move the objects of our cognition closer or further away. Simplicity is also that which filters what is relevant from out of complexity, and complexity is that which enriches simplicity. This is the case with complex systems that arise from the accumulation of simple structures. Such a thought resonated powerfully in ancient atomism, where all qualitative features were thought to result from accumulations of the simplest objects – atoms. Let artificial neural networks simulating their natural counterparts serve as a contemporary example in this respect.

4 The Method of Simplification

Simplification of complex phenomena appears as a fundamental cognitive method, a kind of methodological archetype, both from the perspective of the natural sciences and from that of philosophical reflection. An example of the high level of effectiveness of such an approach is the *divide et impera* ("divide and rule") principle, which is the basis of the most effective computer algorithms, where complex issues are divided into multiple more easily solvable sub-problems. This is also the classical research method recommended by Descartes (2006, p. 17):

> The second was to divide all the difficulties under examination into as many parts as possible, and as many as were required to solve them in the best way. The third was to conduct my thoughts in a given order, beginning with the simplest and most easily understood objects, and gradually ascending, as it were step by step, to the knowledge of the most complex; and positing an order even on those which do not have a natural order of precedence.

Simplification also occurs when we concretize – i.e., move from the abstract, a non-self-sufficient part of an object, to the concrete, a self-sufficient part of it. Recall that the GPT chat in the interview mentioned in Chapter 17 gave an example of a positive feedback loop when pointed in the direction of the creation of filter bubbles: "where people are only exposed to information that reinforces their existing beliefs". Feedback loops can arise between the phenomena being studied and their models: the phenomenon influences the construction of the model, and the model, in response, like a filter, reveals just such-and-such properties of the phenomenon and not others. This is also how the concepts of simplicity and complexity work: they allow one, through feedback, to ascend the spiral of cognition.

5 The Ethics of Openness

In the introduction to this volume, we suggested that openness to the complexity of perspectives and simplicity of expression is a virtue, yet we did not indicate what exactly this virtue pertains to. It is not, after all, a naive openness to all that is complex, nor a simple delight in some engaging expression, even if that expression captures the essence of the phenomenon under study. Virtue lies somewhere in between – between the simplicity of what is vividly felt and the normative complexity of the environment, between what the context imposes and that sort of moral reasoning whose conclusion is the taking of appropriate action.

To act virtuously, as Aristotle (2011, p. 34) taught, means to act as one ought to. Openness to the complexity of perspectives and simplicity of expression is a virtue precisely inasmuch as one ought to be open both to things generally and, where specific things are concerned, to the degree that is right and appropriate. The crown of virtue understood this way – which always lies between extremes – is that *phronesis* mentioned above. It is thanks to it that we know how to find our way about within this tangled multitude of issues that are sometimes simple and sometimes complex. Thanks to practical wisdom, we know when cognition is overly simplistic and when too complicated. Thanks to this virtue, we also know how to choose simple means to achieve complex goals.

6 Art

The inherently simplifying principle of parsimony in thought and action that is widely accepted in science, and whose well-known manifestation is *Ockham's razor*, is also valid in the art world. As Arnheim (2004, p. 59) points out, "the artist must not go beyond what is needed for his purpose". Even seemingly complex masterpieces of art can be characterized by a kind of simplicity, which means that "they organise a wealth of meaning and form in an overall structure that clearly defines the place and function of every detail in the whole" (Arnheim, 2004, p. 60). "In the context of art, the notion of simplicity should be applied not only to the appearance of the work itself but also to the relation between the image seen and the statement it is intended to convey" (Arnheim, 2004, p. 62).

The image is a medium that is not a simple analog of spoken or written language. Therefore, artistic representations are not literal, but rather metaphorical. They are subject to multiple interpretations, affecting us on both intellectual and emotional levels. The perspective of art probably does not in any explicit way bring us closer to giving clear definitions or rigorous descriptions of the relationship between complexity and simplicity. It does, however, allow us to empathize with this relationship, and thanks to that, as a result of dynamic association, enlightenment – itself eagerly awaited by scholars and artists alike – may occur.

References

Aristotle. (2011). *Aristotle's Nicomachean ethics* (translated, with an interpretive essay, notes, and glossary by R. C. Bartlett & S. D. Collins). Chicago, IL: The University of Chicago Press.

Arnheim, R. (2004). *Art and visual perception. A psychology of the creative eye*. Berkeley, Los Angeles, CA, and London: University of California Press.

Descartes, R. (2006). *A discourse on the method of correctly conducting one's reason and seeking truth in the sciences* (translated with an introduction and notes by Ian Maclean). New York, NY: Oxford University Press.

Fleck, L. (1979). *Genesis and development of a scientific fact*. Chicago, IL: The University of Chicago Press. https://press.uchicago.edu/ucp/books/book/chicago/G/bo25676016.html

Hutchins, E. (2012). Concepts in practice as sources of order. *Mind, Culture, and Activity*, *19*(3), 314–323. https://doi.org/10.1080/10749039.2012.694006

Milà, N. C. (2016). On the special relation between proximity and distance in Simmel's forms of association and beyond. In T. Kemple & O. Pyyhtinen (Eds.), *The anthem companion to Georg Simmel* (pp. 81–100). London and New York, NY: Anthem Press.

Nicholas of Cusa. (2000). *The bowling-game* (J. Hopkins, Trans.). Minneapolis, MN: The Arthur J. Banning Press.

Index

actions 2, 9–11, 14–17, 30, 35, 75, 88, 94, 110, 113, 114, 125, 144, 172, 196, 209, 212, 214, 218, 219, 274, 277
adaptive immune system 20, 26, 30
algorithmic capitalism 214
algorithms 75, 120, 121, 206–209, 211, 214–221
analytical 5, 10, 70, 87, 90, 120–123, 129, 135, 137, 140
analytical complexity 123, 124
Anfinsen's dogma 44
antigen 3, 19, 20, 23–26, 30
antigen recognition 3, 19, 20, 25, 26
argumentation 98, 99, 129, 131, 132, 134, 135, 140
artificial intelligence 74, 121, 126, 208, 213–215, 221
artistic expressions 6, 7
asymmetry 213

bioinorganic chemistry 162, 171
biological chemistry 161
broadened mind 16

cancer 35, 38, 39, 107–114, 151
cancer immunity 24
cells 20, 21, 24–27, 29, 30, 34, 35, 37, 45, 47, 49, 51–55, 74, 108–114, 151, 156, 157, 169, 171, 172
character 9–13, 15–17, 72, 92, 103, 120, 122, 125, 151, 177, 181, 185
cognitive 1, 4, 5, 8, 63–65, 67, 68, 71, 76, 78, 82, 88, 98, 105, 155, 177, 183–185, 189, 191, 194, 196, 202, 203, 207–209, 221, 273, 275, 276
combinatorial chemistry 37, 38
communication 1, 3–5, 7, 119–124, 126, 194–196, 198, 199, 202, 214, 218, 221, 227, 230–232, 256
complex 1, 3–6, 10, 16, 19, 20, 23, 24, 27, 30, 35–40, 45, 48, 51–55, 68, 70–72, 78, 81, 82, 87, 88, 92, 93, 103, 105, 108, 111–114, 119–126, 130, 140, 146, 148, 150, 153, 158, 161, 163–165, 167, 175, 177–179, 182–184, 187, 189–191, 194, 195, 198, 201, 203, 208–210, 212, 216, 217, 273–277
complexity 1–8, 10, 14, 15, 34–37, 39, 59, 60, 64, 68, 70, 73, 78, 82, 87–94, 96, 97, 102, 103, 107, 109, 119–124, 126, 129, 130, 140, 145, 146, 177, 179, 188, 193, 199, 200, 202, 206, 208–212, 214, 215, 218–221, 273–277
complexity of communication practice 124, 218
computational model 72, 74, 79, 81
computational psychiatry 4
conditions of conversation 195
content of presentation 176, 177, 184–188
continuity 23, 26, 103, 104, 186, 190, 227, 230
conversation 4, 5, 16, 193–203, 212, 215, 222, 230
copper(I) complexes 4, 107, 109–114
COVID-19 4, 215, 226–232
critical thinking 217, 221, 274

dialogue 7, 12, 13, 17, 197, 199, 201, 202
dichotomous division 99, 100, 104
diegesis 9
discrete 102, 103, 105, 190
disease states 34–36, 38, 52, 82, 150–152, 194
dissonance 60–64, 66–68, 221
dynamic equilibrium 3

electronic referrals 228, 230
emergence 16, 30, 71, 120, 122, 123, 126, 207, 209, 214, 274
emotional experience 10, 11, 15, 274,
emotions 1, 8, 10–17, 88, 199, 211, 219, 274, 277
equilibrium 3, 45, 46, 55

flexibility 3, 10, 45, 51, 123
folding 7, 44–47, 49–52, 54, 164
folding funnel 45
fractal 274, 275
fragment 26, 28, 36, 50, 87, 89, 90, 93, 104, 121, 165, 167, 168, 171, 198, 274

harmonic entropy 63, 64
heat-shock proteins 47–49, 52, 55, 164–166
heterogeneous 108, 114
human diseases 148, 150–155

human health 3–5, 11, 15, 64, 208
hypothesis 26, 27, 29, 44, 64, 76, 198

imagination 5, 10, 13, 14, 17, 87, 97–99, 103, 105, 132, 133, 137, 274
immune system 3, 19, 20, 24, 26, 30
infinity 100–102, 104
instruments 55, 60, 208, 228, 229
integrity 25, 46, 52, 55, 229
invertebrates 26, 29, 30

language 1–3, 74, 105, 120, 136, 194, 195, 197–199, 207–209, 220, 275, 277
Levinthal's paradox 44
logic of conversation 200

mathematics 64, 105
method 9, 10, 12, 14, 16, 24, 52, 81, 90, 91, 99, 100, 104, 145, 147, 172, 196, 209, 229, 231, 276
methodical complexity 2–8, 10, 14
mimesis 9, 10
models 3, 4, 45, 64, 65, 71–79, 81, 82, 87, 90, 100, 102, 103, 122, 126, 148–152, 154–156, 158, 165, 166, 168, 170–172, 182, 185, 186, 191, 208, 209, 227, 231, 232, 276
molecule 24, 25, 38, 39, 45, 51, 114

neurocognitive science 71
neurodegenerative disorders 44, 52

object of presentation 176, 177, 183, 184, 186, 188, 198
objects 3–5, 8, 71, 88, 90, 92, 97, 100, 103, 144, 175–191, 199, 207, 209, 275, 276

partner 10–12, 14, 15, 17, 53
peptides 3, 4, 34–39, 44–48, 50, 55, 107–109, 111–114, 144, 146, 165, 167, 168, 172
performance 7, 11, 81, 82
philosophical discourse 5, 129, 140, 274

primitivism 8, 87, 92
probability 63, 72, 73, 79, 133
protein folding 44, 45, 49, 51, 52, 54
protein quality control 46, 49, 54
proteolytic enzymes 3, 34, 35, 37, 39
proteome 3, 4, 45, 46, 52, 54, 55
proteostasis 3, 4, 46, 49, 55

quantitative complexity 93

RAG genes 25–30
ratio 60–64, 65, 67, 68, 89
receptor 20–24, 26, 53, 78, 108
research methodology 119, 126, 276

scientific 2–4, 8, 72, 104, 148, 149, 150, 152, 154, 155, 161, 162, 182, 191, 202, 209, 212, 274
sequence 3, 21–23, 26–30, 37–39, 44, 46, 54, 74, 120, 136, 137, 143, 145, 146, 154, 165, 167–170, 187, 188, 196
simplicity 1–8, 10, 12, 15, 35, 39, 59–65, 67, 68, 70, 81, 82, 86–94, 96, 129, 130, 134, 140, 142, 145, 146, 161, 194, 200, 202, 203, 206, 220, 221, 262, 273–277
stability 3, 26, 60, 64, 65, 67, 143–147, 168
Stanislavsky system 10, 14
Stanislavsky's method 12, 16
stoic 5, 129, 130, 133, 177
stoic philosophy 130
substrate specificity 35–39

T-cell receptor 20–25
teleconsultations 227, 230, 233, 234
telemedicine 227, 231, 232
theory of everything 59
theory of types of dimension 177, 186–188
therapeutic task 71, 72
truth 5, 15, 89, 162, 182, 198, 201, 206–210, 213–216, 218–221

verbal complexity 5, 195